Telecommunications Billing

A.T. Bell

"Telecommunications Billing," by A.T. Bell. ISBN 1-58939-737-1

Library of Congress Control Number: 2005904922

Published 2005 by Virtualbookworm.com Publishing Inc., P.O. Box 9949, College Station, TX 77842, US.

Manufactured in the United States of America.

Table of Contents

Telecommunications Billing

Telecommunications History

In the Beginning

Telecommunications technology started to evolve early in the nineteenth century. This was a time when inventors were just beginning to discover the advantages of using electricity in their inventions. No longer were these newer, evolving technologies, being expressed entirely using mechanical principles. Instead, devices developed to improve communications were being created which now combined both electrical and mechanical technology.

Electricity, at the time, was being transformed from a somewhat entertaining curiosity to an intense driving force within our civilization. The first form of communications to utilize this technology, in this early historical period, was the "telegraph." Many people are fairly familiar with this device from the old black and white war movies seen on TV. Ships in trouble would survive by calling for help using communications equipment, such as the telegraph, to send "*Save Our Ship* " (S.O.S.) messages using Morse code. After the initial invention, these devices soon became heavily commercialized. Moreover, they began to be extensively used for communications by the military and railway stations.

As encountered with other inventions, there were many discrete steps and contributions to its creation. Though, many times, we are not always aware of how many people contributed to the device. Instead, we only tend to remember who was credited for the final accomplishment. When the device reaches a state where it can become commercialized, the person who made the final contribution is typically rewarded with credit for the entire discovery. This often is not the actual inventor of the technology, but someone who has improved upon the accomplishments of others.

One of the first primitive telegraph devices to receive attention was initially presented to the public in 1819. It was demonstrated by an inventor named Hans Christian Oersted. He first demonstrated the device at the University of Copenhagen. In his presentation, he showed that a compass needle could easily be deflected by a small electrical current. This first demonstration was a good example of the very basic electrical theory which was used by the first telegraph inventors. The magnetic field represented how electricity could be used to move a mechanical element. This was an important discovery.

Not soon after, an inventor from the United States, named Joseph Henry, implemented a similar signaling device in 1831. It further expanded upon the initial accomplishment that Oersted had achieved. However, his invention was a more practical one. It was biased more towards use in communications applications. It consisted of a horizontal magnetized steel bar supported on a pivot. When a nearby magnet was energized by an electrical current, one end of the bar was forced to strike a bell, which consequently conveyed a signal.

This was another major milestone and a step in the right direction, but it was not until the year 1832 that more complex electrical circuits began to be introduced. It was then that a Russian diplomat, named Paul Schilling, improved upon Henry's work. He again proved that the deflection of a compass needle could be moved by an electrical current. But, this time, he also proved it was also the starting point for a basic electrical telegraph. He found that when electric current flowed through a coil of wire surrounding the needle, it would then deflect. This gave people some initial insight as to how electricity affected the needle movement. Also, by adjusting the electrical field, the movement was able to change. This was a pivotal moment in history. The door was now unlocked and waiting for someone to step through.

It was soon after that a now famous, early inventor began experimenting with different communications codes. He first reviewed the other inventor's approaches to telegraphy. Then, after some further experimentation, he was able to develop a simple code which he felt would allow people to easily converse electronically. He had foreseen these simple codes as an initial means to provide a short hand notation for this new evolving form of communications. In his mind, it would eventually allow the telegraph to be used for communicating between parties over long distances. The inventor's name was Samuel F. B. Morse. His name eventually became widely associated with the invention of telegraphy. In fact, today the famous communications code still bares his name.

It was Morse who had initially realized the potential of the telegraph. He saw beyond his peers. He saw the art of application. He was able to anticipate how it could be used and the possibilities which existed. He saw this potential as early as 1832. To further pursue these ideas he worked on representing some important telegraphy concepts. By 1835, he finished constructing the first mocked-up representation of his thoughts. Then, convinced he had come up with a feasible communications code solution for telegraphy, he moved forward and applied for a patent in 1837.

In 1844 he gave the first full-scale demonstration of the working device. It was then that he communicated the now famous message, "What hath God wrought? ", between Washington, D.C. and Baltimore, Maryland. Soon after, in 1846, the Western Union Telegraph Company was founded. They pushed the business world into wide spread adoption of this early technology. As such, the first commercial implementation of Morse code was adopted by this company. It was promoted as the standard method of transmitting coded messages using the telegraph.

About the same time that telegraph technology was beginning to flourish in the United States, in England it was becoming an interesting topic of conversations. The word was out and the public was starting to see the potential of this new exciting technology.

It was in England that a man named Charles Wheatstone had successfully invented, what at the time someone would consider, a more advanced telegraph. It was an electronic version of the telegraph which was designed for commercialization. He established an English version of the device which could be adopted by the masses. Not long after his invention became known by the public, he was credited for the invention. It was referred to as the "electric" telegraph. By mid-century the telegraph circuit had become an essential part of the Industrial Revolution. While in the beginning, Wheatstone's fame was initially little or nil, the telegraph did well and was soon adopted for use by the railroad, military, and the press.

In its earliest states of commercial use, the telegraph was only being found exclusively at railway stations. It was initially engaged in the routing and scheduling of trains. Even as it became a popular topic for newspaper articles, the telegraph soon was one of the primary methods of rapid communication for the newspapers themselves. To further promote the rapid communications capabilities of the telegraph, newspapers began to jump on the bandwagon with respect to reputation for fast communication. For example, it was commonly used to advertise that a paper had the most up-to-date information. Many papers included the word telegraph in their titles, for example, "The Daily Telegraph" in London. The world soon took a giant leap forward due to the advances in communication initiated by the electric telegraph.

Invention of the Telephone

Yes, the telegraph was an essential invention. However, the telephone built further upon this technology and was considered an even larger milestone towards mass communications. The story behind the invention of the telephone begins in 1875. This is a time when the telegraph and the Atlantic cable were mere technological wonders.

It was then, that Alexander Graham Bell, a successful Professor of Elocution, was diligently working on the telephone invention. People really don't hear the term Elocution much anymore, but it was a common term in those days. Elocution was considered an art form in a sense. It was considered an expert manner of speaking involved with controlling one's voice. But it referred; not only to one's voice, but also to the gestures they would perform while speaking. It involved the full gambit of conversing. It was concerned with the mannerisms as well as the presentation of one's self during a conversation. These both were considered important elements. They were intensely studied at the time to depict how someone was to correctly convey information.

This professor was not only teaching at the time, he was also engrossed in an invention. The physical attributes of the device could only be described as a crude type of harmonica. It was comprised of a clock spring reed and a magnet with a wire attached. Using these elements he attempted to get sound from one end of the wire to the other. He had been working on the invention for three years. It was in June of 1875 when the invention finally yielded the results Bell was searching for.

Alexander Graham Bell

Alexander was a successful teacher of speech and acoustics. He also considered himself a novice in electrical theory, but was by no means, fluent in the use of this technology. It was in 1875 that he eventually achieved his success. It was then that without the use of a battery, with no more electric current than was made by a couple of magnets, that he was able to achieve his goal. The sound waves generated by the reed had, themselves, been successfully carried along a wire. Then, at the other end, they were changed back to their original tone.

Though, even after the first initial sound came from the device, it was more than forty weeks before the telephone could do more than make strange speechless noises. These were new concepts at the time. Bell was still learning how to better control his invention. It wasn't until March 10, 1876 that it finally spoke words and was considered a somewhat practical device. It said distinctly, "MR. WATSON, COME HERE, I WANT YOU."

On his twenty-ninth birthday, Bell received his first patent for his invention. The patent number was 174,465. This was the most valuable single communications patent ever issued. He had created something so new, at that time; there was not even a name for it in any of the world's languages. In describing it to the officials of the Patent Office, he called it "an improvement in telegraphy," when in truth, it really wasn't. Other early inventors experimenting with telephone technology had always worked from the standpoint of the telegraph, but most never came close to achieving what Bell did.

Alexander worked from the standpoint of the human voice. He merged the two sciences of acoustics and electricity to make his invention a success. His study of "Visible Speech" had helped immensely. Ultimately, it had trained his mind in such a way that he was successful in his accomplishment. He felt he could almost mentally see the shape of a word as he spoke it. At the time, he knew quite a bit about speech and considered how words and vibrations acted upon the air. As a basis for early telephone theory, he thought of how vibrations were carried from the lips to the ear. His family was heavily involved in the teaching of speech. He was merely a third generation specialist in this topic. He believed that for the transmission of spoken words there must be "a pulsatory action of the electric current which is the exact equivalent of the aerial impulses."

Thanks to his efforts, Alexander was eventually credited not only for the telephone invention, but in the end, for taking preliminary steps towards starting the largest telecommunications corporation in the world. He never could have predicted that the system would grow to include such an array of organizations which now bear his name. The collection of the companies formed, based on his invention, were initially referred to as being part of the "Bell System."

Though Alexander was extremely interested in communications technology, he was not interested in big business or the concepts of free enterprise. He considered himself to be dedicated more towards technological inventions and humanitarian endeavors. Though, once the telecommunication system concepts were proven, he still continued to give many speeches on various aspects of the invention. After all, he still wanted to promote the device. He began to see the industry driving towards a telecommunication infrastructure, which used the telephone as a key component. It was then that he became merely a stockholder in the company. But, even then, he never stopped continuing to publicize his creation.

The only tangible assets of the Bell Patent Association included some early Bell Patents. Including, "Improvements in Transmitters and Receivers for Electric Telegraph," his telephone patent, No. 174,465, and "Improvement in Telegraphy" (March 7, 1876). Two additional patents regarding this technology followed these as well.

The Bell Telephone Company

In 1877 Alexander Graham Bell was married to the love of his life, Mabel Hubbard. This was his business partners' daughter. It was not long after the wedding that he and his partners conferred to consider forming a new telecommunications company. The company would utilize entirely telephone technology as a method of communication. At the time, they felt this would be the best way to promote this new technology and help it gain acceptance. All of his business associates agreed. They created history by forming the *Bell Telephone Company* (BTC).

After the BTC was formed in 1887, the company set

out to revolutionize the telecommunications industry. They started with just 5,000 shares of public stock. The initial shareholders of the newly formed company consisted of the initial business partners and their close relatives. This is how the shares of the new company were initially divided:

Name	Shares
Alexander Graham Bell	*10*
Mabel Bell	*1497*
Gardiner Hubbard	*1387*
Gertrude Hubbard	*100*
Thomas Sanders	*1497*
Thomas Watson	*499*
C. E. Hubbard	*10*

Table 1.0 Bell Telephone Company Share Distribution

The telephone was now working well. The bugs had been worked out of the device. The new company was established and working diligently to look for ways to promote the technology. Even now, more than ever, the device needed to be further introduced to the business community.

The first individual to undertake this task was Gardiner G. Hubbard. Gardiner was Mabel's father, now Bells father-in-law. He was the first person to really successfully advance the telephone business. He had been working effortlessly to campaign for improvements in telegraphy. His first step toward capturing the attention of an indifferent nation was to beat the big drum of publicity. He saw that this new idea of telephoning must be exposed more to the public. As a result, whenever he had the opportunity, he openly talked about the telephone and attempted to sell people on the idea of using the telephone for mass communications.

Further, to run an effective telephone publicity campaign, Hubbard asked Bell to perform a series of, then considered, sensational feats with the telephone. Even after some of the initial demonstrations, there were still a large percentage of people who denied that spoken words could be transmitted by a wire. When Watson, Bells assistant, talked to Bell at public demonstrations, there were newspaper editors who publicly were still skeptical. So, once and for all to convince these doubters, Bell and Watson planned an astounding test for the telephone. For their demonstration, they first borrowed the telegraph line between Boston and the Cambridge Observatory. They then attached a working telephone to each end. Then Bell stayed at one end while Watson stayed at the other. They began to talk to one another. For the first time in history a SUSTAINED conversation by telephone happened! It lasted for three hours. Each one taking careful notes of what the other was saying and what was heard.

After this accomplishment, the publicity regarding the telephone soared. Immediately following this feat Bell arranged a series of ten lectures. He was paid a hundred dollars a lecture. Not bad money in those days. This was actually the first money he had ever received for his invention. His first lecture in the series was in Salem before an audience of five hundred people. A pole was set up at the front of the hall supporting the end of a telegraph wire that ran from Salem to Boston. Watson then sent messages from Boston to various members of the audience. An account of this lecture was sent by telephone to The Boston Globe, which announced the next morning "This special dispatch of the Globe has been transmitted by telephone in the presence of twenty people, who have thus been witnesses to a feat never before attempted."

Alexander's presentation in Salem convinced the newspaper editors. For the first time they began to believe in the telephone. Greatly encouraged, Hubbard and Alexander prepared a little circular which was the first advertisement of the telephone business. It is an oddly simple little document today, but to the 1877 mind it was startling. It modestly claimed that a telephone was superior to a telegraph for three distinct reasons:

- *No skilled operator is required, but direct communication may be had by speech without the intervention of a third person*
- *The communication is much more rapid, the average number of words transmitted in a minute by the Morse sounder being from fifteen to twenty, by telephone from one to two hundred.*
- *No expense is required, either for its operation or repair. It needs no battery and has no complicated machinery. It is unsurpassed for economy and simplicity.*

Hubbard decided that the time had come to further organize the business, so he created a simple agreement which he called the *Bell Telephone Association* (BTA). This agreement gave Bell, Hubbard and Sanders a three-tenths interest each in the patents, and Watson received one-tenth.

However, even though the telephone was now starting to gain popularity, as for being in a position to start a business, there was no capital. NONE! The four men by this time had, what would be considered, an absolute monopoly of the telephone business. But even after several successful demonstrations, the only man willing to stake money on the future of the telephone was a man named Thomas Sanders. He ran a business which cut out soles of shoes for shoe manufacturers. The first five thousand telephones were made using his money. It was many expensive months before any monetary relief came to Sanders.

Alexander believed that they had something terrific. He was convinced that someone, somewhere, would be willing to purchase and help evolve this technology. But until this good individual arrived, they could do no more than struggle ahead, and acquire whatever customers they could. That is, as long as the customers were close and it was cheap to provide service to them. Bell was still continuing to perform somewhat articulate presentations. He painted mental pictures of a universal telephone service to applauding audiences. Mr. Sanders was continuing to lease telephones anywhere it was feasible to do so.

The Competition

Western Union was the major opposition to telephone technology. They had by now completely swallowed most of their competitors. They were focused on monopolizing all methods of communication by wire. This worried Sanders. He felt, at the time, that the Western Union was the only company which could promote the telephone. The only hope that Sanders saw was if Western Union might agree to purchase the Bell patents. He, along with his partner Hubbard, offered the rights associated with the telephone, to Western Union for

$100,000. Western Union refused. They didn't believe the telephone would be a good investment decision.

But besides the operation of its own wires, Western Union was supplying customers with various kinds of printing telegraphs and dial telegraphs, some of which could transmit sixty words a minute. These accurate instruments, it believed, could never be displaced by the telephone. They continued to believe this until one of its own subsidiary companies, "Gold and Stock," reported that several of its machines had been superseded by telephones.

In response to this news it did not take long for Western Union to begin pursuing telephone technology. It took action by organizing the *American Speaking Telephone Company* (ASTC) with $300,000 capital. With all of its great wealth and status, it set out to trample upon Bell's patent. At the time, and to the complete incomprehension of Bell, it serenely announced that it had "the only original telephone," and that it was ready to supply "superior telephones with all the latest improvements made by the original inventors; Dolbear, Gray, and Edison."

However, this new interest in the telephone by Western Union had an unexpected turn of events. The Bell group, instead of being driven from the field, was almost immediately lifted to a higher level in the business world. It was then that the telephone ceased to be a "scientific toy," and became an article of commerce. It began, for the first time, to be taken seriously. Western Union, in the endeavor to protect its own private lines, became involuntarily a motivator to lead capitalists in the direction of the telephone.

Capitalists immediately came to the financial rescue of Bell and Sanders. They started investing money in the Bell patents. Two months after Western Union had given its endorsement to the telephone, these capitalists organized a separate telephone company to do business in New England. They put fifty thousand dollars in its treasury.

It was not long after that Sanders business partner Hubbard found himself leasing telephones at the rate of a thousand a month. Now he and Sanders were slowly no longer acting as promoters. They were, instead, acting as General Managers of the business. Soon extremely basic telephone exchanges were being started in a dozen or more cities.

Growing the Business

Sanders realized at this point that they needed someone with more business experience. They wanted an individual with a good track record. They wanted someone who could manage the business in its currently accelerated rate of growth. The new position of General

Manager was eventually offered to Theodore N. Vail for a rate of $3500 per year.

A mere one week later the young man took his seat as General Manager in a tiny office in Reade Street, New York. As a result, the evolution of the industry began. The arrival of Vail at this critical moment emphasized the fact that Bell was now ready to make significant progress in building the infrastructure. The new General Manager had, of course, no experience in the telephone business. Neither had any one else. But he, like Bell, came to his task with great enthusiasm. As a boy, he played around the first telegraph line, and learned to put messages on the wire. His favorite toy was a little telegraph that he constructed for himself. Then, at twenty-two, he went west in the vague hope of possessing a bonanza farm. But it didn't take long before he swung back into telegraphy. In a few short years he found himself working for the Government Mail Service in our nation's capital, Washington D. C.

By 1876, he was at the head of this service, which, in turn, he completely reorganized. He introduced the bag system in postal cars. He made a valiant effort to wage war on waste and inefficiencies in the mail system. By virtue of this position, he was the one man in the United States who had a comprehensive view of all railways and telegraphs. He was the perfect candidate for developing a national telephone system.

He had the needed experience for straightening out the tangled affairs of the telephone. Line by line, he mapped out a method, a policy, and eventually a complete system. He introduced a larger view of the telephone business. He even persuaded half a dozen of his post office friends to buy stock in the telephone business. In less than two months the first Bell Telephone Company was organized, with $450,000 capital and a service of twelve thousand telephones.

Vail proceeded to start building a corporate business policy. He stiffened up the contracts and made them good for only five years. He confined each agent to one place, and reserved all rights to connect one city with another. He established a department to collect and protect any new inventions that concerned the telephone. He agreed to take part of the royalties in stock, when any local company needed to pay for service this way. He took steps toward standardizing all telephonic apparatus by controlling the factories that made it.

These various measures were part of Vail's plan to create a *National Telephone System* (NTS). His objective was not the mere leasing of telephones, but rather the creation of a federal company. This federal company would be a permanent partner to the Bell business to assist with management of the telephone infrastructure.

However, Western Union was still fighting to be the predominate force in the telecom industry. They threw

Bell managers into distress by introducing the Edison transmitter. It was superior in capabilities to the telephone. The five months that followed were the darkest days in the childhood of the telephone.

How to compete with Western Union, which had this superior transmitter, a host of agents, a network of wires, forty million in capital, and a first claim upon all newspapers, hotels, railroads, and rights of way. These were the immediate problems that confronted the new General Manager. In the effort to conciliate a hostile public, telephone rates everywhere were low. Hubbard had set a low price of twenty dollars per year. This cost covered the use of two telephones on a private line. When the exchanges became widely accepted, the rate was even lower. It seldom was more than three dollars a month. In the larger cities, where Western Union had the most influence, the lives of the telephone pioneers became difficult and were sometimes met with hardships.

In Philadelphia, for example, a man named Thomas E. Cornish was attacked attempting to establish the first telephone service. No one would, officially, grant him a permit to string wires. His workmen were even arrested when they attempted to do so. Proponents of the telegraph warned him that he must either quit, or be driven out. When he asked capitalists for money to get the lines strung, they replied with a resounding no. Finally, he was compelled to resort to a different strategy.

He needed some backing in order to accomplish his goal. He needed someone with authority to help. Colonel Thomas Scott, the President of the Pennsylvania Railroad, assisted with his efforts. He wanted a line ran between the railroad and his house. As soon as Cornish had put this line in place, he kept his men at work stringing other lines. When the police interfered, he showed them Colonel Scott's signature. They then left him alone. He used this method to put up fifteen wires, before the trick was discovered. It was not long after that, with only eight subscribers, he founded the first Philadelphia exchange.

As you can imagine, such a battle did not put much money into the pocket of the parent company. The letters written by Sanders, at this time, prove that the business was still in rough shape. Month after month, the Bell Company lived from hand to mouth. No salaries were paid in full . Often, they were not paid at all.

Then, as if things could get worse, Bell returned from England. He went there in an attempt to further expand the business. He announced that he was broke. He also announced that he had failed to establish a telephone business in England. In fact, he said he somehow must acquire a thousand dollars to pay his immediate debts. By this time, he was thoroughly discouraged.

Fortunately, he received a letter from Francis Blake, a man from Boston. Blake told Bell that he had invented a transmitter as good Western Unions. He agreed to sell the invention to Bell for stock instead of cash. The possession of his transmitter instantly put the Bell Company on an even footing with Western Union. It was greatly encouraging to the capitalists who had invested money, and it stirred many others to come forward.

The general business situation had by this time become more settled. In only four months the company had twenty-two thousand telephones in use. As a result, they reorganized into the National Bell Telephone Company, with 850, 000 dollars in capital.

When it came to long distance, the infrastructure was still not up to par by 1880. In order to expedite the availability of long distance, a new telecom organization was established, named American Bell (AB). They purchased quite a few manufacturing installations to produce the telecommunications equipment the company needed. However, even with this, by 1882, the demand was so significant that American Bell acquired the Chicago based electrical manufacturing firm, Western Electric. They gave them exclusive rights to manufacture Bell telephone equipment.

By 1876, each individual's telephone line was connected directly to another individual user. In 1877, a switchboard was installed in Boston so each phone could be connected to a switchboard for routing calls vs. being connected directly. This was the beginning of the mass consumer use of the telephone. The infrastructure was now established.

Chapter

2

Introduction to Telephony

Basic Telephony

Let's review the basic telephony components. It is important to not only be introduced to the concepts, but to understand how these elements work together. Moreover, how do these elements fit into today's telecommunications infrastructure. As a first step, we must consider all of the elements which comprise a basic phone system. The most notable element is the telephone itself.

The telephone introduces us to a starting point for voice transmission. It incorporates some very basic, but important, communications concepts. These concepts encompass advanced voice to electronic signaling conversion techniques. This conversion is necessary to transform the voice into a form where it can be sent across a wire. That is, assuming another transmission material such as glass is not used. Once the signal reaches the other end, it is decoded into the original voice signal.

The telephone system is comprised of a transmitter, receiver, and a transmission medium to carry the voice signal from one end to the other. The telephone is the ultimate communications device. It incorporates two of the three required telephone system components. It acts as both transmitter and receiver.

It is a significant telecom device indeed. It uses an apparatus which converts mechanical vibrations to an electronic signal. This device is similar to a microphone. So, when speaking into the phone directly, the voice is converted into its electronic equivalent.

Today, the technology has continued to advance beyond what Bell accomplished. The features of today's telephone appear endless. From caller id to call waiting, the device is continuing to supply and support more features to meet consumer demands.

Most phones today create an analog electronic signal which is derived from the voice. An analog signal is the same as what your house power uses. It is a signal which alternates from one magnitude to another at a certain rate. For your house, this is 60 cycles per second. For your voice, it is much higher and varies in rate significantly. From this, a digital signal representation is first produced before transmitting it across the wire. The phone can not only transmit these signals, but also act as a receiver and decoder for these signals. Once decoded, it will allow someone to hear the original voice transmission.

Local Loop

From a basic electrical circuit perspective, a customer's telephone is connected to the "LOOP." This is the transmission area between an individual's telephone and the nearest "*Central Office*" (CO). The CO contains the switches which route phone calls across wires from the origination phone to the destination phone. The wires are a paired cable (two copper wires twisted together). This wire is usually used in one of four gauges: 19, 22, 24 or 26. These cables are then placed into "binder groups."

A binder group consists of multiple wire sets. Each color coded set consists of 25 twisted pair copper wires. These wire sets are covered by a protective sheath. The group is used as a transmission medium for both, voice and data.

The telecom business term "Local Loop" refers to the telecommunications circuit comprised of the central office, telephone, and the transmission medium between these entities. Local loop facilities, sometime referred to as "The Last Mile," are a critical and somewhat capital-intensive network component. They are used in delivering reliable and adequate telecommunications services to a local carrier's customers. In today's competitive markets, "Local Loop" facilities must be cost-effective, reliable and capable of accommodating changing customer needs.

Central Office

A consumer of telecommunication services utilizes a telephone to communicate with other parties. Both, the origination and the destination telephone are connected to a central office. The "*Central Office*" (CO) is the switch exchange which contains the intelligence to route calls to their destination. It holds the business rules for moving calls across switches and trunks through central offices and local regions. When calls are sent from one central office to another, the call must go across these trunks.

Trunks hold many copper wires. They are used to transport the communication signal between both calling parties. The transmission medium used to connect a customer's phone to the *central office* (CO) is generally composed of copper wire.

Eventually, it will be optical cable. Optical cable provides transmission capabilities through light instead of an electronic signal. This allows for a greater signal "bandwidth" to be available during a transmission. This means a greater number of simultaneous calls can exist on one fiber strand vs. its copper counterpart. Also, transmission distance is significantly increased through the use of these fibers, while noise on the line is decreased.

Trunks

A trunk can be defined as the communications medium which connects one central office to another. They are focused communication channels which carry traffic between switching systems. Each call on the trunk can use the bandwidth available for the duration the call. After the call gets to the central office, it goes to the *Main Distributing Frame* (MDF). This is where the loop ends at the CO in the "Local Loop" configuration. It is the end point but where calls are then directed to another CO and set of switches depending on the call destination.

The calls go into "Feeder Routes" where multiple "Local Loops" converge onto the MDF. The Loop on the other end of the circuit is called the *Subscriber Network Interface* (SNI). This is where the "Loop" ends on a customer's premises. This is the other side of the loop from the MDF.

The essential elements of your telephone are as follows:

Switch Hook	This device is typically where your phone handset rests and is used to provide network control signals to the Central Office.
Dial	Your phone either uses a rotary dial system or a DTMF (Dual Tone Multi Frequency) system. So respectively you are sending a series of pulses or tones.
Transmitter	Converts your voice to electrical signals and transmits them over a transmission medium.
Receiver	Receives incoming electrical signals and decodes them back into sound.
Ringer	Alerts individuals their is an incoming call.

Table 2.0 Telephone hardware elements

Key Set

Other than basic telephones, many smaller businesses have also incorporated something called "Key Telephone Systems" to handle multiple incoming lines.

Included in such a system is a device called a "key set." A key set has the same functions as a telephone except it has even greater capabilities.

Usually, businesses establish this phone system with both, telephones and key sets. The capabilities of this phone system configuration are catered towards business customers. It allows an organization to manage calls more efficiently. For example, with a key set, the user can put a call on hold while answering or making another call. They have access to more than one line at a time, and they can answer the same line from more than one key set device. Typically, a key set system is comprised of the following basic elements:

KSU	This is an interface between the lines from the Key Set System and the Central Office.
Key Sets	Key Sets have a hold button and line appears buttons representing all

	of the lines which the user has access.
On Premises	The wiring on the customer's premises which permit the correct lines to appear with the right key set.

Table 2.1 Key Set elements

Each customer which incorporates a "Key Telephone System" will generally have a phone switch residing on their premises. These phones, as well as residential phones, are referred to as *Customer Premises Equipment* (CPE). These switches are usually one of the following types:

- *Private Branch Exchanges (PBX)*
- *Private Automatic Branch Exchanges (PABX)*
- *Automatic Call Distributors (ACD)*

PBX

These switches are connected to the *Central Office* (CO) through trunks, specifically designed for this purpose. The type of switch used is dependant upon the business customers needs. A PBX or PABX switch is typically a small switch that is located between the user's telecom network and the central office. Today, when organizations need to connect a phone system to a telephone exchange, they must utilize a *Private Branch Exchange* (PBX.). A PBX is a telephone system which is private to the organization. It is used to connect them to the public switched telephone network.

Another switching system is called "Centrex," is similar. It has become generally referred to as the "Central Exchange." However, instead of being located at the customer's location, this switching equipment is located in the local phone companies CO. However, it still performs the same switching functions as the PBX.

ACD

What is an *Automatic Call Distributor* (ACD)? We've all encountered the use of voice queuing systems. These are generally referred to as *Integrated Voice Response* (IVR) or *Voice Response Units* (VRU) systems. It's difficult to reach a live person sometimes when requesting support or placing orders with an organization over the phone. This is exactly what an ACD does. It is a switch that queues calls to specific telephone sets. Your call waits in the queue until a live person can answer it. Most provider call centers utilize an ACD to route calls to company representatives.

Digital Technology

In the beginning of telecommunications technology, the telephone network was originally designed to carry voice traffic only. However, due to the evolving nature of computers, the infrastructure now carries voice as well as data across its network. Digital technology has slowly been introduced. Today, after many infrastructure upgrades, much of the telecom network hardware and telecommunications customer premises equipment are now comprised of digital technology.

Direct Inward Dialing

Along with these digital technologies, another major innovation towards automation of organizational communication was *"Direct-Inward Dialing"* (DID). DID was a significant innovation. It allows incoming calls to an organization to be routed to a PBX or a key system without human intervention. The organization must purchase software with groups of special telephone numbers which operate across multiple trunks. The central office looks at the number dialed and determines which business area the call belongs to and routes it appropriately. This is very similar functionality to what a telephone operator would do at the phone company except in a faster more automated fashion.

Voice Band

To fully understand all of the processes and technologies associated with the telecom empire, we must learn some of the basic concepts of telephony prior to moving forward with discussing the billing aspects or leading technologies prevalent in the industry today.

One of the most important concepts to be familiar with in communication is the concept of bandwidth. Many of us have been witness to consumer media reports regarding newer technologies which require a lot of bandwidth. Usually, the report is related to some type of newer automation or communications technology which incorporates video or high speed data.

Bandwidth determines how much electronic or optical "signal" information can be transmitted over a transmission medium. Signals provide a representation of the original video, data, or even voice. Mediums, such as copper wire or optical cable, have significantly different physical properties. These properties, themselves, determine the amount of bandwidth available. If a transmission medium has more bandwidth available, it can transmit more information. An optical cable can provide a significantly larger bandwidth capability than its copper counterpart.

When someone speaks into a telephone the human voice varies in frequency from 50 to 20,000Hz. However, most of the intelligible information is understood between 300 to 3300 Hz. This is the "Voice Band." The "Voice Band" has an identifiable bandwidth. This bandwidth can be calculated by taking the difference between 3300 and 300 Hz. If we calculate the resultant bandwidth in this fashion, we can determine for the basic "voice band" it is 3000 Hz. This is what we'd think our transmission medium needs to support. However, the voice information by itself has a 3 KHz bandwidth. But beyond the voice information, there are typically some "out of band" analog or digital control signals associated with the elements of your phone, so the bandwidth of the "voice band" is allocated at 4 KHz.

Analog vs. Digital

What is the difference between digital and analog communication signals? Digital electronic signals are a series of electronic pulses. Analog signals are not. An example of an analog signal would be the music a home radio plays. An example of a digital signal would be what your home computer uses to process instructions.

An electronic waveform can be represented visually, for the purposes of analysis or discussion. Special electronic test equipment called Oscilloscopes can be used to view these waveforms. If we were visually representing these different signals, what would they look like? Digital pulses would be represented as a series of square waveforms. They would vary in magnitude, but they would be flat on top and bottom. An analog signal would be represented by a smooth, magnitude alternating, sinusoidal waveform. They would not be flat at the top or the bottom. Instead, they would represent a smooth transition as the magnitude of the signal varied.

Digital refers to everything being represented by discrete binary electronic signal pulses representing either a 1 or a 0. With a digital pulse, information has only two states. Like a switch, it is either on or off. For example, an on state represents a 1; an off state represents a 0.

A sinusoidal analog waveform can be converted to a digital representation. This can be done by taking the sinusoidal information and identifying discrete amplitude thresholds. The thresholds identify where the signal needs to be converted to a digital 0 or a 1. Today it has become relatively easy to electronically translate any analog waveform to a digital form.

In communication systems, after an analog to digital signal conversion, a digital pulse now represents the original voice. This conversion process is through the use of special analog to digital conversion technology. Integrated circuits, which are small fingertip sized electronics, are used to perform the necessary signal translations.

Signal Impairments

Many times limitations in the phone system are due to signal impairment. Signal impairment is the degradation of a signal over a transmission medium. It can be due to many causes, most notably, attenuation, noise, crosstalk, or signal echo. These signal impairments are common for both, analog and digital transmissions. Attenuation refers to the reduction of amplitude during the transmission of a signal from one point to another. All signals attenuate due to the properties of the transmission medium.

One significant limitation to a communications system is noise. Typically, either impulse or random, noise interferes with a telecommunications system. Impulse noise is an unexpected spike in amplitude (voltage magnitude) of a signal. Random noise is more of a constant noise. These can be caused by a number of factors. Examples are the environment, insulation sheathing around the transmission medium, or electrical equipment being used for transmission etc.

This noise can sometimes jump onto a signal and affect the information being carried. Sometimes, it can even render the signal unusable. The ratio of the signal to noise, can affect the ability of the receiving equipment to correctly decode the message. Use of filtering and low noise amplifiers in electronic instrumentation, can sometimes assist with reducing this level.

Chapter

3

Introduction to Billing

Telephony Billing Systems

Almost every telecom service provider incorporates a telephony billing system. These systems transform service usage into monetary compensation. They also provide ways to manage customer account information. They perform credit scoring, rate available products, accept payments, and reconcile credit and debit information with third parties. In a nutshell, telecom billing can be defined as the process which refers to how billable charges are collected, processed, and invoiced.

For most people the interaction with their phone company is infrequent, except for receipt of their phone bill. They typically only contact their provider for a small number of reasons. Maybe, they want to acquire a new service or product. It possible they need to disconnect an existing service. Or third, and most common, is the need to transfer service from one address to another. For all of these services, the customer receives an invoice. The invoice comes at regular intervals, sometimes referred to as "bill periods." This invoice contains all of their usage, recurring, and one time charges. It includes any taxes and fees associated with services, provided for a given time period. These fees can be anything from hardware fee usage, to local, state, and federal taxes.

In order for a provider to be able to generate an invoice, they must collect billable charges from the various telecom network or service elements. The network events which generate these charges are first collected, and then used to rate service usage.

Billable Charges

Billable charges may also come from third parties or affiliations which are providing the service. These charges may come in the form of usage charges, recurring charges, or taxes and fees. The charges are associated to a particular customer. The customer is associated with their account information and billable charges. Billable charges, typically, come in some type of notification event for the billing system. For example, a third party such as a clearinghouse may send an event. This event indicates what charges are applicable for a customer within a given time period.

The billing system uses this event information to initiate business processes to properly rate and invoice the customer. Typically, the billing system would respond to multiple billing events. It would aggregate and categorize the resultant information. The billing system would use this information to generate the customer's phone bill.

Billing/Working Telephone Numbers

A customer's association with their provider needs to be identified. This comes in the form of some type of IT system identifier which the provider can use to lookup customer information. For Telco types of services, the primary identifier for a provider to recognize a customer's account information is generally through their *Billing Telephone Number* (BTN). This may or may not be the phone number the customer uses on a day to day basis. The phone number the customer uses for placing calls is typically called the *Working Telephone Number* (WTN). This may be one and the same as the BTN. The customer may actually have many phone numbers with various services from their provider existing on different phone lines. However, the customer's billable charges are generally aggregated and associated with a BTN in the provider's IT systems. Then the BTN will cross reference to the WTN.

Voice Services

Basic voice services are usually referred to as *Plain Old Telephone Service* (POTS). Business and individual residential customers, both, utilize voice services. To subscribe to these services, customers must utilize the expertise of a telecom local service provider. It may be an *Incumbent Local Exchange Carrier* (ILEC) or a *Competitive Local Exchange carrier* (CLEC).

The billing system associated with services can be complex. It may be comprised of multiple transactional islands of information pathways to connect IT systems. It may use extremely large *information technology* (IT) data warehouses to store information. It may manage various connection points and initiate events to other IT systems, within the provider's enterprise.

An ILEC, has information technology professionals within the service provider's organization. These people help develop, and maintain, robust information systems. These billing IT systems handle all billing operations, billing consolidation, electronic payment, and remittance operations.

North American Numbering Plan

If the customer lives in the United States, Canada, Bermuda, or Puerto Rico they are using the *North American Numbering Plan* (NANP). The NANP format consists of a 10 digit phone number. The first three digits of a phone number is considered the *Numbering Plan Area* (NPA). This represents the customer's area code, and is unique to a particular calling area usually defined within a particular state. Local calls, generally, share the same NPA as the originating phone number. So, if a customer is calling from a 314 area code to another phone number with this same area code, it is almost always a local call from an ILEC or CLEC billing perspective. The service provider manages local calling areas based on this identifier as well as *Local Access Transport Areas* (LATAs).

The FCC defines a LATA as an area with a community of interest. Essentially, a LATA will consist of a region and can serve one or more area codes. Generally, smaller lesser populated states, will have only one LATA. Other denser populated areas, may have multiple LATAs associated to them. One example is the state of California. It has a high population density with currently 11 LATAs. LATAs themselves are managed by the FCC and not at a state level. When calling across LATAs, you will encounter various toll charges between each region. In dense regions, one area code can be a part of multiple LATAs. A call originating from one calling region, and terminating in another, is considered an "interlata" call. These calls will many times be long distance calls. However, they also may be in-state calls or inter-state calls. A LATA may occupy an area in more than one state. An "intra-LATA" call is a call placed within the same calling region. Intra-LATA calls may be local calls or they may be long distance calls, depending on your local telephone company.

When reviewing your phone number, the second set of three digits in your phone number is considered the number exchange (NXX). This number identifies the ILEC receiving your inbound call. The last four digits of your phone number represent the extension number. This extension number is unique to your particular phone line. The phone bill itself, is essentially an invoice for services provided. Key items to note on the customer's phone bill, are recurring and usage charges, taxes and fees associated with hardware or governments, and the billing cycle, for a given bill period.

Billing Re-Engineering

If there is one word which explains what a provider needs to continually do in order to remain competitive today, it's INTEGRATION! As providers launch a new generation of services, they need to take seriously the challenge of upgrading their existing billing processes to adequately accommodate future needs. Through acquisitions and technology upgrades, it has become a race to see who will offer the next generation services first.

Upgrading existing billing systems forces providers to incur a significant expenditure. However, it is important that providers review their IT system needs and make these critical investment decisions. Not only in terms of their expected capital costs, but costs for next generation services and support systems. In addition, costs for operational monetary expenditures needed to improve efficiency and increase overall customer satisfaction.

One technology that has assisted with the evolution of telecom billing systems is “e-commerce.” E-commerce technology has now matured and business processes related to this technology are under critical review and undertaking core process re-engineering efforts. Questions are emerging regarding how to automatically bill, and receive payment from, business customers. How do we do this in the most efficient way possible? How do we integrate the billing process, with other internal telecom processes and with business partner IT systems?

As more and more consolidation takes place, today’s telecommunication industry must continually make compromises. They must cooperate with business partners. They must also work together with the competition in order to support a mutually beneficial future for both parties. When a provider considers what is needed in the area of billing, across business partners in this complex FCC regulatory environment, they must utilize billing systems which are flexible. In order to do this, they must use standard technologies, and adhere to fixed business rules. These systems must be built to account for future needs. They must allow future functionality to be easily integrated into the architecture.

Common Business Frameworks

It is important to address these difficult industry challenges. If providers have redundancies in their IT systems, they must embrace centralization of common business rules. They must develop common IT software interfaces and application frameworks, upon which, new billing systems can be built. It is important to streamline operations and minimize business process redundancies. Also, crucial to a providers' survival, given industry consolidation, is business knowledge. It is important to recognize, and take advantage of intellectual capital, across the business.

Newer billing business systems must support the needs of the customer, through creation of consolidated business processes. It is important to make a customer's life easier and not harder. Newer processes must take these customer care concerns into account. For example, let's say a customer has multiple products on their account. These products are requested from business partnerships with the provider's business affiliations. The customer does not want to receive a separate bill for each affiliate product. They want one single integrated bill.

Billing processes are rapidly changing due to the various billing options available today. From land-lines to wireless services, the realm of billing is becoming vast. Integration of systems and telecommunication network services is needed to assure flexibility within the next generation billing architecture.

It is clear that this diversity of billing opportunities will need seamless operation across business affiliations. They will need the flexibility to handle prepaid, postpaid and real-time billing based on a wide range of metrics. These may include network service measurable items, such as volume, content, duration and timeliness across trunks and systems.

Billing services are now being provided which utilize batch feed techniques to send billing information between providers. These services provide information, not generally available in the provider's IT systems, where the service was ordered. For example, information feeds may be used to enable one aggregated bill presentation to the customer, representing products across affiliations. It may be used to reconcile account information between each providers' IT system. This exchanging of information, allows providers to present features like integrated bills, and wholesale billing arrangements. It also, many times, provides benefits from a support perspective when something goes awry.

Convergent Billing

Convergent billing can correctly be defined as the integration of all service charges, onto one billable invoice. This allows providers to provide a unified view of the customer's charges. This centralized focus relating to one's phone charges, is becoming mainstream within the telecommunications industry. More and more, providers are taking this approach to assure customer satisfaction.

Within the highly competitive marketplace of phone service, convergent billing is another way to lure and keep important customers. Both CLECS and LECS, understand the huge value to an organization, by offering such services. It, essentially, allows one stop shopping for multiple products and services, supplied by just one provider. One bill for multiple services greatly simplifies the charging process for customers, making it an incentive to stay with one provider.

Customers prefer a single number to call for billing support. They do not want to call multiple providers or multiple support areas for billing questions relating to different services. They, also, want their phone bill charges to represent product offers made to them by the customer service representative, during the order process.

This can be a difficult element in billing when the products or services reside across business partners. When an order is initially processed for a product or service by a provider, it must be distributed to the appropriate IT system for processing. This may be an internal system which is Telco specific. It may also be a business partner, which has a wholesale *Service Level Agreement* (SLA), to provide a service for the Telco. The affiliate system may then price or rate the product being ordered (pricing refers to the product offer where rating refers to the off the shelf individual price). This forces an affiliate system to send reconciliation information from their billing system to the Telco, so that the invoice can be generated from one billing system, which can aggregate all of the charges and generate "one" invoice.

There is always a certain amount latency which is encountered as part of this process. So, though the order may be placed on Monday, it may be Thursday before reconciliation can take place. This sometimes causes the phone bill to represent a prorated bill, which does not accurately and exactly match, the offer the CSR provided. Many times, it is only a portion of the bill which causes the customer confusion.

One important service the provider performs is the abstraction and convergence of billable charge information. They receive this charge information from many different information sources. Providers might need to integrate with business partners for product, account, and rating information. A billing system must be able to collect and aggregate this information to be used for the final invoice creation.

One of the most difficult aspects of working with business partners and their billing systems, is alignment of billing cycles. This is important so that the customer sees a bill which reflects what they expect. They expect to see exactly what they were quoted by the *Customer Service Representative* (CSR) that sold them the product. Many times due to pro-ration of services, disparate system feeds, and different billing cycles across affiliate products, this can be very difficult.

Account Information

Billing systems can be extremely complicated. They house sometimes millions of phone records with countless financial usage and tax calculations associated to each one. The information displayed on a customer's phone bill actually represents information from many billing processes. A customer's account is generally represented in the form of a *Customer Service Record* (CSR). The CSR provides information regarding all of the products and services on a customer's account.

When a customer requests a new service from their provider, a new updated CSR is automatically generated. The records are maintained by an information technology system, called an *Operational Support System* (OSS). The services a customer orders are referenced by the OSS using a *Uniform Service Order Code* (USOC) and *Field Identifier* (FID) code. USOCs and FIDs are how service providers track products and their features in a somewhat universal way.

When ordering services, the OSS will generally generate a customer number which is unique to the customer and provides an added layer of security. When a customer has provisioned network services, they can also be identified to a provider by their phone number, or some other unique network address element. The service itself can sometimes be used to determine the identification of the customer. Sometimes, however, when new services span across affiliations, regions, or across multiple accounts, a unique customer number must exist to identify all of the services and accounts a customer has. A common identifier can be used to group the account information across businesses.

Due to countless acquisitions of smaller service providers by larger organizations, business systems have now evolved to represent a customer in a different way. This has caused many data integration difficulties with matching customers between providers. For example, John Doe in the providers IT systems needs to also exist in a business partners IT system. There may be even be multiple John Does. They need to be identified uniquely.

Perhaps one option would be a number such as their *Social Security Number* (SSN). Another may be a combination of known customer information. A problem arises, however, when the number on one system doesn't match the number on the business partners systems for the same customer.

The solution is generally a customer repository or data warehouse which expresses these relationships has become common place. It aggregates this information. It allows for a centralized reference point for customer information across all business affiliates.

If we review the *Customer Service Record* (CSR), it is important to consider the properties on this record associated to a customer. The format of the CSR itself is typically in a line-item format that is relatively easy to use. Two kinds of charges are recorded onto the customer's phone bill from the CSR, these being monthly recurring charges and service connection charges.

Some key elements which exist on the customer's CSR are as follows:

Account Number

The Account Number consists of the NPA, NXX, and the last 4 digits of the customer's telephone number. In addition, a CUSTOMER CODE is usually associated with this number as well.

Directory

Number indicating which Yellow or White Pages Directory your number resides in.

Service Order Number

This number is generated when a customer orders services and they are provisioned.

Connection Date

Date services are enabled

Activity

Detailed service usage information

Telephone Number

This number consists of the NPA, NXX, and 4 generated digits. Together they represent the customer's phone number.

Description

Describes details of the service

USOC

Universal Service Order Codes represent a product or service.

Monthly Rate

Presents the recurring rate for one or more products or services.

FCC Charges

Defined subsidies the FCC allows the

service provider to collect from the customer.

Class of Service

Type of Service

Billing Name and Address

Billing Information

Service Connection Charges

Charges collected for new installations or connections.

Notes

General information about your account

Products and Services

Detailed costs associated with each provisioned product or service.

Amount

Detailed Charges

Activity Date

When service was established

Taxes

Tax types associated with provisioned services.

Listings

Indicates whether you phone number is listed in the white pages, yellow pages, etc.

Information Services

Used for special aux services such as DSL or other.

These CSR attributes associated with the customer's phone bill are crucial to gaining an understanding of the billing process. It determines how a phone service customer is charged for services.

In addition to general phone service billing, the *Federal Communications Commission* (FCC) has mandated that all telecommunication carriers, whose facilities are used to provide long distance services, are entitled to a share of the compensation paid for that service. This revenue helps providers to provide facilities, over which the long distance company, can provide calls to its customers.

When AT&T was forced to divest itself into different operating telephone companies, the nationwide telephone network was effectively divided into two basic types of components: Local and Long Distance. So after divestiture, a caller making a long distance call used facilities belonging to at least two different companies: one providing local service and one supplying long distance facilities. This division of facilities has created the need for compensation, so that each company providing facilities for a call can be compensated for the use of facilities used to provide the call. In telephony, the Last Mile - the local loop - consists of the provision of service from the end-user customer to the *Point of Interconnection* (POI) of the *Long Distance Company* (LDC). In the USA, this segment of service provision is called Carrier Access. The Last Mile portion of a call addresses how the local service provider is compensated by the LDC, for supplying the circuit and completing the call, to the actual customer or end user of service. At both ends of the call, the originating local carrier and the terminating local carrier have a stake and an obligation in the provisioning of the call. Therefore, compensation is due to each. Both the access billing process itself, and the tools for tracking and calculating the bills, are known as *Carrier Access Billing* (CABS).

Before divestiture in 1984, a flat amount per call was allotted to the Non-Bell companies as part of their settlement for provision of service to the end-user. This flat compensation did not take into consideration the type of facilities provisioned to the specific end-user. It did not take into account, the distance the traffic had to travel to the POI, the length of the call, and/or the route traveled. Because divestiture invalidated the previous settlement structure, and because of the increasing number and complexity of the services that *Local Exchange Carriers* (LECS) provide, there had to be a way of appropriately compensating the local telephone company for providing facilities to the user.

Thus, as part of the Modified Final Judgment of 1983, the Federal Communications Commission mandated access charges. One of those access charges is the *Carrier Access Charge* (CAC) to long distance companies that connect to the *Local Exchange Carrier* (LEC) network. This charge enables local telephone companies to partially recover the cost of the "Local Loop," which refers to the outside telephone wires, underground conduit, telephone poles, and other facilities that link each telephone customer to the telephone network. (LECs also assess flat monthly charges to the end users.)

The original model for an access tariff was written by a group of staff from AT&T. It used criteria such as switching, the traffic route, tandem functions, and minutes of use into consideration for local compensation, as well as other, more technical data. The "Access" documents were written and revised a number of times before their final release in 1983. On January 1, 1984, the right to bill for access was mandated.

Both, the originator of the call and the terminator of the call, are entitled to compensation. In some cases, there is more than one entity providing the service to the customer. In the case of CLECs, there may be a hand-off to the Bell tandem. In this case, both the CLEC and the Bell Company require compensation. This type of service is more complex for the long distance company to validate, as it probably will receive separate bills from the

Bell and the CLEC that are billed at different times of the month and aggregated onto bills.

Billing Management

Competition

Today telecommunications providers are at a turning point. Intense competition from de-regulation has resulted in price pressure and decreased revenues in land-line voice services. Telecommunications providers need to cut unnecessary expenditures, retain customers, and determine business strategies to increase revenues to stay competitive.

They need to review expenditures and minimize investments in older technologies. They need to monetarily embrace driving their business towards newer maturing technologies and revamp their existing billing systems and telecom networks.

Architecture

Before anyone can begin to effectively manage billing operations within a telecom organization, they must have a comprehensive understanding of the complexities associated with the billing processes. They must not only know the federal regulations, taxation, and compliance issues from a business perspective, they must familiarize themselves with the engineering processes which comprise essential billing operations. From a billing business perspective, providers have many important processes and IT System interactions to consider. Business processes such as offer management, sales negotiation, billing, ordering, networking etc. all play an important role in effective billing.

An *Operational Support System* (OSS) is the basis for a provider's billing process support structure. These IT systems are the legacy software driven information systems commonly used *by Incumbent Local Exchange Carriers* (ILEC). These systems are the backbone of the telecommunications infrastructure. They provide core operations for the telecom business. This includes network enablement. They cover the full gambit, from provisioning of telecommunication services to billing.

Since divestiture, these systems have evolved across many different telecom companies into dissimilar software implementations. For example, ordering systems and billing systems, which were once identical, pre-divestiture with AT & T, are now entirely different business systems with different functionality, different software interfaces, etc. This has enabled these companies to work productively in a self contained environment. They previously did not need to worry about anther providers IT systems. After all, at this time, they did not need to interface with another providers billing or ordering OSS implementation. That is until now. As new acquisitions have taken place in the telecommunications industry, the need has arisen for these systems to collaborate, integrate, and share their information to meet the needs of the providers business.

This important point becomes even clearer as companies realize that assurance of revenue and customer retention is through the offering of packaged products and services. Customers are less likely to switch providers when they have multiple products through one carrier. The other important consideration regarding the support structure is the business itself. This includes the synchronization of data between different IT business systems. It includes creation of new manual business processes. It also includes new automated business workflow processes which need to interact with backend *Operational Support Systems* (OSS).

These information system processes and the software applications associated with them are referred to as

Business Support Systems (BSS). These are systems that have been designed to effectively manage workflow, integration, and business processes throughout a telecom organization.

Today, many years after divestiture, the telecoms legacy OSS infrastructure can still be characterized as extensively using proprietary technologies. This ranges from the network elements themselves, through the OSS/BSS and to the *Enterprise Application Integration* (EAI) software architectures which need to be employed across a provider's enterprise to effectively link these elements together. These existing OSS architectures have become very costly to maintain and are currently inflexible to market demands.

Business Model

With unregulated newer and cheaper technologies at the doorstep, providers need to be able to develop a

"Business Model" which allows them to effectively compete. They need to determine how to align their business model to embrace these newer technologies, and at the same time, utilize *Customer Relationship Management* (CRM) techniques to hold onto their existing customer base.

These strategies need to take into account the future trends within telecommunications and prevent customers from flip flopping between providers. A successful telecom business model needs to address changes to both, billing and customer care business processes, with respect to the changing marketplace.

The future of telecom is moving towards high speed Data, Video, VoIP and WIFI. Developing a business model which meets the criteria to be successful can be challenging to say the least. This book reviews billing strategies which any company can perform to stay competitive.

Global Resources

Continued advances in technology, access to "Global Resources," and refined business practices, have reshaped the way telecom service providers work with their customers. Long gone are the days of turning over complete control of customer support functions to a third-party and hoping for the best. Instead, through the use of well designed business processes, historical trending analysis, and a focused strategy for customer retention, a provider can be assured customers will be less likely to switch providers.

Intense pricing structures has encouraged providers to many times embrace practices such as outsourcing their entire IT or customer service staff to third world countries such as India or the Philippines. Though sometimes a controversial subject, the most costly resource of a business is its people. If a provider can get the same quality of work cheaper, it can be tempting to embrace this approach whole heartedly. The important consideration a provider must make is which areas of the business is this feasible; which areas will this actually increase the overall time to market and discourage customers from working with the provider. This is especially important in such a critical area of the business such as billing.

As a general guideline, non-mission critical initiatives are candidates for outsourcing. Mission critical initiatives and initiatives where time to market is a primary consideration are not. Through communication barriers, time differences, and multiple channels of leadership, software development using offshore resources may cause many cost overruns and ultimately cost a provider millions in potential lost revenue. Time to market for new services is crucial given the fierce competition today. The key is to streamline processes, eliminate unnecessary overhead, and increase synergies between business groups where possible.

Functional Resources

Organizational change is many times the key to remaining competitive. A provider's objective should be to align organizational elements with existing company strategies and established priorities. The organization must be flexible enough to reconfigure itself based on these goals. This can be through development of a strong matrix type of organizational structure where specialized functional resources are allocated to projects.

This involves having specialized work groups such as billing experts or software experts, fluent in a particular technology, be allocated to the project. In this situation, resources are specialized in a particular discipline. For example, a provider may have a billing expert, a rating expert, and a customer care expert. Providers may also have software resources, to be used for initiatives where customer custom software is needed, to automate business processes. These resources may be fluent in project management, software development, or a particular technology such as web development. The goal is to develop experts. Not a broad spectrum of knowledge in multiple disciplines per resource. Instead, people who are masters in a particular area.

Resource Pooling

The other avenue to pursue is the concept of resource pooling. Resource pooling involves development of a centralized pool of resources with a broad spectrum of skills. These resources can be allocated to a variety of initiatives allowing new teams to be formed quickly. The difficult part of resource pooling is establishing a pool of resources with enough experience and knowledge to be effective. It is important that the learning curve on new initiatives be minimal. These resources need to be subject matter experts in the discipline where they are providing value.

Elimination of redundant efforts within a provider's organization can also be a successful strategy in beating the competition. If the provider has multiple sales channels with parallel software initiatives and identical business processes, then a strategy for development of common services and processes which all sales channels can leverage should be considered.

Planning

One important aspect of managing initiatives for the telecom business is proper planning. Planning is essential for driving initiatives within the business to completion. Sometimes, it may seem overwhelming given the dimensions involved. The key, as with any large initiative, is to break the task down into small and more manageable pieces.

In its simplest form, a plan is essentially an entity which has a collection of distinct actions, initiated incrementally, by a sequence of events. Many times, we have dependencies between these events, which force us to look at how we implement our plan. A plan is dependant upon a set of actions, which has multiple relationships between the plan and the actions themselves. The actions are dependant on a relationship of multiple consequences, which also have multiple relationships between a set of actions.

Now I know attempting to understand these relationships, and attempting to verbally express them, can be daunting to say the least. You may sound like you've delivered a Zen riddle to your listening audience. But understanding the relationships and dependencies between actions and consequences within a plan is essential to being able to manage it effectively.

Billing

Telecom billing systems are used as the centralized point at which all telecom charges, whether they be fixed, recurring, or usage based, will converge and be used to create the customer's phone bill. These are comprised of the IT systems and business processes which rate products and services. They aggregate billing account information and generate the final invoice for the customer.

Unlike telecom services in the past, today it is common for a service provider to offer a vast assortment of services. Through the use of business alliances, they can provide many products and services which lure potential customers. They commonly decide to offer packages which include products from both the provider's organization and a business partner's organization.

Today, many providers have affiliations with wireless providers, television satellite providers, as well as other telecom related services. They may also provide service for services such as high speed internet and data. As a result, billing systems have become highly complex and require many business process steps to be completed due to the amount of external systems involved. A great deal of the complexity exists due to the need for cross-affiliate billing. For example, one provider sells a product which belongs to a business partner. There must be an automated way of reconciling these charges and getting them to be distributed on "one" phone bill.

This may sound simple. However, a primary consideration is that affiliate organizations, many times, use different technology and business systems to process their billing charges. The billing charge, itself, needs to be registered in both billing information systems. It needs to be processed in such a way that the billing cycle lines up with the customer's expectations. What they are charged for in a given month should line up with their order. Customers don't want to see a charge on their phone bill at an unexpected time. They want to see what they expected for charges in a given month. They typically also don't want to see multiple bills. They want to see one bill with all products listed across many affiliates and have those charges reconcile.

Due to these different billing cycles between affiliate organizations, reconciliation can become quite a challenge. Consider if a product being ordered is part of a package. The customer wants to save money by buying a group of products vs. ordering the products individually. They order a package with one Telco product, one wireless product, and one video product. When they receive their bill they expect the charges to be consistent with what they ordered. They don't want to see partial charges just because an affiliates billing cycle is different than that of the provider. They want one consistent invoice with all incurred charges at one time.

A telecom billing system is comprised of "Collection," "Rating," and "Invoicing" business processes. Billing software is what is used to automate these business processes, and it is a crucial aspect of any successful billing system.

Input, Output, and Processing

These business processes are comprised of basic processing elements such as input, output, and the processing in between. The "input" is the collection of "Billable Charges" such as "Usage Charges," which are time sensitive charges. The input can also be wholesale charges from third parties, recurring charges, or one time retail charges. They can also be federal, state and local government tax and fee charges.

The "Bill Processing" is what actually aggregates information from these different input sources. This process determines any additional tax and fee charges which may need to be added to the bill. It creates the billing records in the proper format for creating the phone bill. In order to bill correctly, the billing system must have a way of getting the dialed number and determining what rate is associated with it. The system will need to get the locality name, determine the duration of the call, and store the cost associated with each call. This is important if the customer is to be invoiced properly.

The "Output" process involves putting phone billing records into a format that can be used for printing. Many times, the billing records are forwarded to a third party for the actual printing and delivery process.

Billing software provides a way to enter rate information into the billing system. Rates refer to the pricing of various product offerings the provider is promoting. Rates are generally location specific. They are associated with the customers *Telephone Number* (TN) or *Trunk Group* (TG), in the case of Telco billing. For other telecom network services such as VOIP, if usage charges were to apply, the IP addresses would be used to determine origination and destination of call charges. Essentially, a TN or IP Address is considered a network addresses, with respect to the telecom network. Location information usually exists in the telecom IT billing systems. It presents some level of granularity to represent a location. It may be an address, region, state, or country, for example. It may represent a general or specific location where the service is being provided. Sometimes, this constitutes a relationship to a mapping to a city, state, or region. It is associated with the customer's service address or associated with their account information.

Wholesale vs. Retail

Billing systems can be comprised of "Wholesale" or "Retail" business arrangements. For example, as a consumer, you may order DSL from an *Internet Service Provider* (ISP) and they have a wholesale arrangement with a telecom provider to provide the DSL service. The service is purchased from the provider at a wholesale rate. Then the ISP adds an additional charge for the service and offers it at a retail rate to the consumer. This requires the ISP to submit the order to the provider. The provider then needs to provision the service on their network and add the transaction to their billing system.

Generally, the customer information will be added to the billing system when the account is created. It will be flagged, in some way, indicating it is a wholesale arrangement through a third party such as an ISP. Many times, high speed internet services may be through a provider affiliation. In this case, billing reconciliation must occur between the provider and the affiliation offering the high speed internet service. This is, in addition, to reconciliation between the provider and the ISP.

Billing Systems

Intro to Billing Systems

Telecom billing systems are generally comprised of the following business elements:

- Customer account information
- Invoicing engine for bill generation
- Rating engine for determining product rates, taxes and fees.
- Product marketing interfaces for defining products and promotions
- Consumer ordering business processes
- Telecom or affiliate networking hardware

Each of these elements plays an important role in any telecom business billing process. Customers have a distinct relationship to the provider through their billing account information. This information is used to maintain what products and services the customer has ordered and is used to bill the customer. This is also what a *Customer Service Representative* (CSR) in the provider's "Call Center" would use to manage customer information. When a customer orders products and services, this information is updated in a near real-time fashion.

Every billing system generally has an "Invoicing Engine" which is used to generate the customer's telecom bill. It uses the account information to determine what products the customer has existing on their account. An invoicing engine takes into account the aggregation of billing content, tax and fee calculations, and bill formatting.

The "Rating Engine" is used to detect usage and the rates involved with particular network elements based on the customer's rate plan. This includes determining tax and fee location information and max available usage, in the form of "prepaid" types of operations. A rating engine, generally, uses some method of storing the rate information for ad-hoc retrieval during the rating process. This is, many times, through an IT system or a table in a database. This rating storage mechanism maintains a relationship between a rate and a plan. It is then listed on the customer's account information, so the billing system knows how to charge the customer for network usage sensitive charges.

A "Product Marketing" group generally creates product definitions and promotional offers to be used by the billing system. This group may make changes to rate plans or promotional offers at any time. Typically, it is on a recurring schedule, or when promotions are only offered for a specified duration. When the duration is over, the group determines a strategy on what to offer next. This is determined by many important factors. These factors include what the demand is at a given time for certain products, competition, network capacity, and supply.

Consumer ordering business processes are used to actually order and provision the service the customer is requesting. These processes include using ordering and billing software to manage the customer's account information. These processes exist in the provider's "Customer Care" department, generally. This is the process where the customer is requesting service and the service is billed and provisioned (enabled) on the provider's network.

Telecom networking hardware is used to enable the service the customer is requesting. The term is used to refer to the backbone of the telecommunications or affiliate service infrastructure. This includes various telephone switches, transmission mediums such as copper and fiber, as well as specialized *Customer Premise Equipment* (CPE), which is many times required for various services.

Billing System Types

There are, generally, two types of billing systems which exist in the telecom world. "Post-Paid" billing services and "Pre-Paid" billing systems. For post-paid services, "Collection Software" exists at the switch in the *Central Office* (CO) which collects data and builds a *Call Detail Record* (CDR). The CDR includes call specific information, such as call origination and destination. The CDR is then used to build the invoice which is sent to the customer.

For pre-paid types of billing systems, software exists at the CO which is used to decrement the account, based on usage. Both, of these systems utilize a process which matches the customer's rate plans against their service. Through the use of a real-time rating engine, individual calls are rated based on the plan, specified on the customer's account. When service usage spans across multiple provider's networks, CDRs are reconciled through the use of third-party "Clearinghouses."

Electronic Bill Presentation and Payment

More and more providers are offering services for online billing and presentation. This, significantly, reduces the costs associated with rendering, printing, and distributing the bill. Depending on the billing complexity, some providers require software to be distributed to the customer, while others only require a browser.

Customer billing software, whether it is desktop or browser based, generally, provides functionality for the customer to pay their bills online and see a copy of their latest invoice. Many times, a billing arrangement can be established using the software, which provides automatic payment from the customer's bank or other financial institution, to the provider. This provides a convenience for the customer and also a benefit for the provider.

Software used for billing large businesses are, generally, more feature enhanced than software used for individual residence customers. Many times, the customer is provided functionality to generate reports based on their billable charges. If the billing spans across multiple regional locations or branches, the software, many times, will allow the user to break down the charges to a more granular level. For example, many times they can generate reports for a human resources department of the business or a specified regional office.

Billable Charges

When considering strictly network usage types of services such as Telco voice or data services, the "Billing Process" involves receiving *Usage Detail Records* (UDR) from various telecom network hardware and software elements. These usage records only include specified usage information on network elements, such as a particular switch in a central office, at a particular location. The billing rates, themselves, are used in calculating what is needed for the customer's invoice. This is greatly influenced by location and the customer's rate plan. They are associated with each UDR, in the sense that the rating engine uses rate information associated with a particular rate plan. The rate plan is identified on the customer's account. The charges are calculated and aggregated periodically to be used with *Call Detail Records* (CDR) to eventually produce invoices.

The invoices are sent to the customer on a given "Billing Cycle" (recurring time period). When the payments are received from the customer, they are "Posted" to the provider's billing systems. Usage charges are one set of charges which begin after services are ordered from a provider. These, and one time charges, are enabled through the use of "Ordering Software."

Ordering Software is used to take the orders from a customer and process them into the backend billing and provisioning systems. This process is referred to as a *Service Negotiation Session* (SNS). This usually occurs by calling the provider's call center, contacting the provider over the internet, or through an *Interactive Voice Response* (IVR) system. Provisioning systems are IT systems which, actually, enable a product or service on the telecom network. They include all IT service ordering operational support systems. If business partners are involved in processing an order, the affiliate billing systems must generally send the billable charge information back to the Telco Billing System. This, usually, happens after any services have been provisioned by the affiliate. This is so the telecom provider can place the charge on the bill which goes to the customer. It is used to reconcile any payments between the provider and the affiliate.

In order to understand core billing processes, it is important to understand the complex issues related to ordering and billing reconciliation.

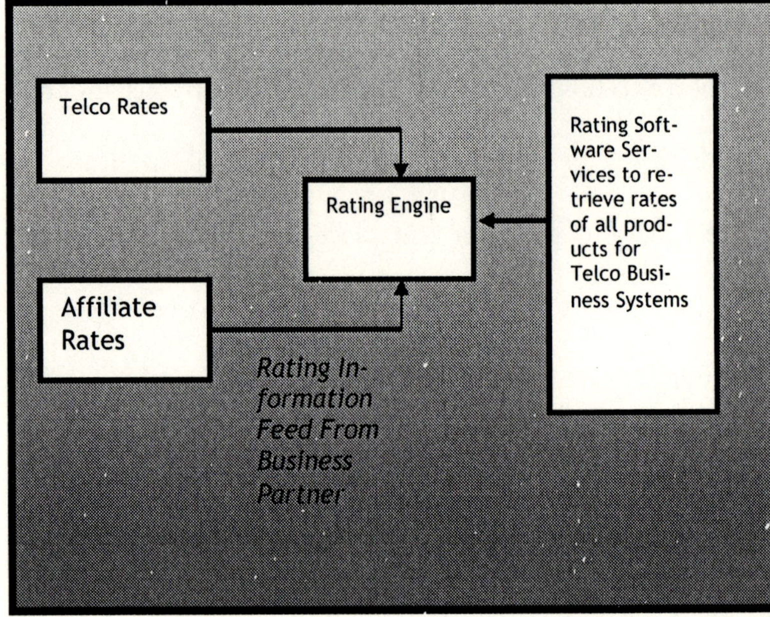

Figure 5.0 Rating engines used to rate products and services

Obtaining Product and Rate Information

One important consideration when rating products is how a provider obtains product rate information into the telecom billing systems from other business partners. For example, if the business has affiliations with other providers, such as a third party wireless provider, how do they receive the rating information they need. How do they effectively rate their packages, which may include a wireless product? One approach providers take is to mediate information from affiliate sources. They do this through the use of one or more rating systems. These are used to consolidate rating information across business partners. Generally, when many companies are involved with different billing systems, products are almost always rated in different ways. If the item being billed for is a network service, such as a Telco voice product, then location information places a decisive factor in what the product will cost in a given location.

Not only does the provider need to mediate rating information, they need a way for ordering processes to access the information. This is through the use of rating software. This rating software has interfaces into it allowing for other business systems to access the rate information for different products. The typical billing process involves collecting usage information from network equipment (such as switches), formatting the usage information into records that a billing system can understand, transferring these records to a rating engine, which assigns charges to each record, receive and record payments from customers, and create invoices through an invoicing engine.

Product Ideation

Concepts for new products can come from a variety of sources. These include employees in the provider's business, competing companies, sales leads, market demands etc. Once an idea emerges, the ramifications must be well thought out. Then, feasibility studies must be performed, to determine whether creation of the product is worthwhile. It is important that the new product be aligned with the strategic goals of the provider. Defining and creation of new products should be a quick and simple process, to assure the best time to market.

New Connect or Service Change

When a customer is asking for new service or a change in service, they contact one of the provider's sales channels such as a call center. The customer then converses with a *Customer Service Representative* (CSR) to establish a new account, or modify the services on an existing account. The CSR enters the customer's service orders into the system, checks for credit worthiness, then provisions and bills the customer for services. If the service is phone related, the customer is provided with a phone number so that the customer may start using their service once it is provisioned.

The phone number is derived from a pool of available phone numbers. This process of managing these pools of numbers is referred to as *Network Address Management* (NAM). Network address management refers to the ability to reserve a number or network address from a pool of numbers. These numbers represent a telephone number. For example, a *Customer Service Representative* (CSR) in a call center would use ordering software to order a new service. If the customer was requesting a new "Land-Line," the CSR would be presented with a list of available telephone numbers. The CSR would then select a number and it would be reserved for this *Service Negotiation Session* (SNS). The management and reservation of these numbers is crucial during the SNS. It will be the number which identifies the customer's account or a particular service on the account.

For example, typically in a call center, CSRs will use a customer's telephone number and state to look up the customer's billing account records. It will use these values as identifying traits for a particular service.

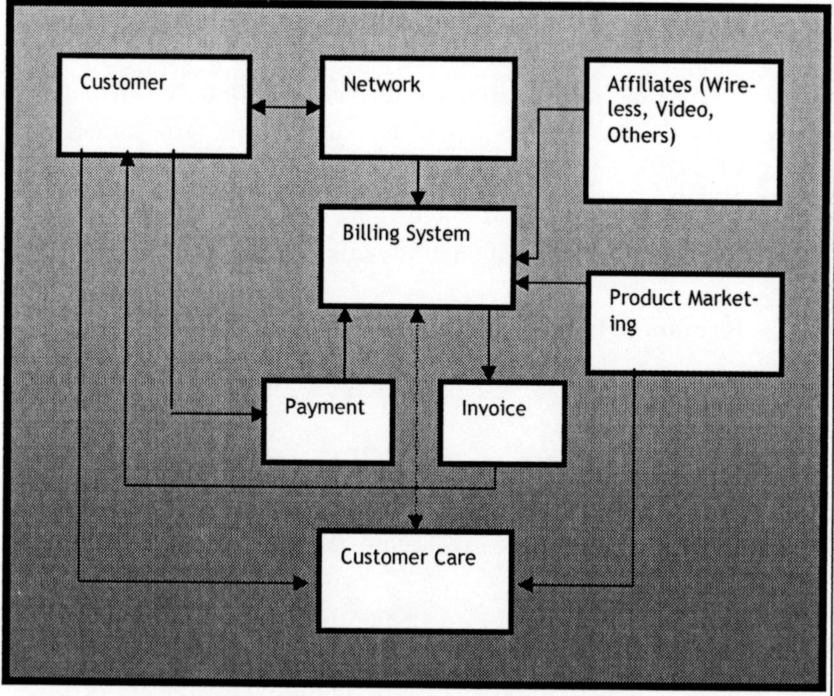

Figure 5.1 Customer to billing system interactions

Basic Billing System Interactions

The basic standard telecom billing system is shown above. It expresses the relationships between the various business organizations. It expresses interactions between the customer and the provider. As shown above, the customer can call for support, make payments, as well as receive invoices and network services.

Usually, when a customer uses a service, usage information is recorded at the *Public Switched Telephone Network* (PSTN) gateway. If the service is a prepaid type of service, then the gateway communicates with the rating engine to calculate the max available time limits. This is based on usage and the remaining balance in the customers account available for the service.

Let's say a customer prepaid for 100 minutes of wireless service. The rating tables in the rating engine (used to store product rate information) would indicate what the rate would be per minute for this type of service. Then, it would record the time the individual was on the call and determine the available time left that the person would be allowed to stay on the network. As a result of service usage, billing systems have software and optional hardware components that receive *Usage Data Records* (UDR). These records are received from various telecom network elements. They are then validated, rated, aggregated, formatted and presented on an invoice that is sent to the customer. Events may be received from many sources. After receipt, they must be converted into a standard format that is recognizable by the carrier's internal billing processes.

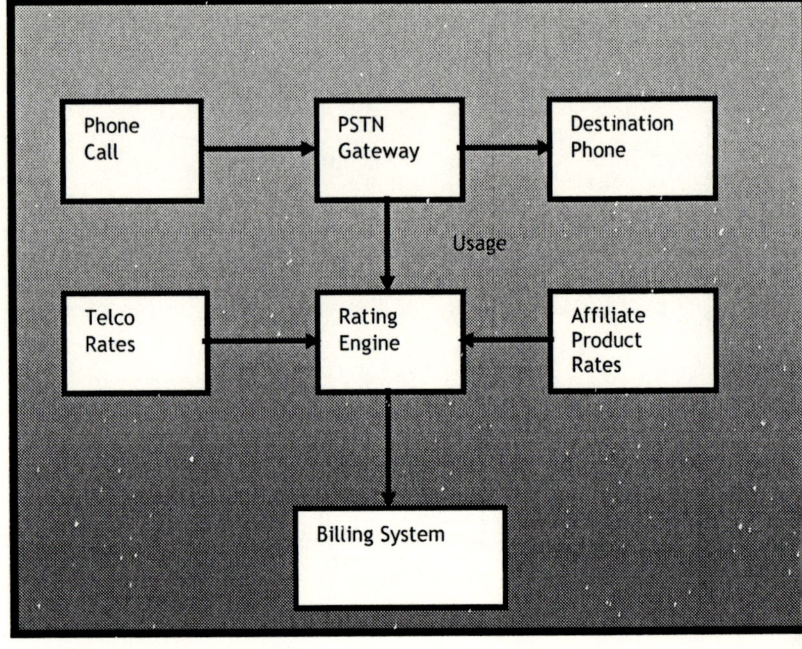

Figure 5.2 Real time rating used to rate network voice services

Rating Collection

The above diagram shows how real time rating is performed within a telecom billing system. The PSTN gateway records usage and passes on this information to the rating system. The rating system determines the amount of time left based on the customer's billing arrangement. Then usage information is received and billing charges are calculated for the call.

The most essential aspects of any billing system are the rating engine and the businesses processes which generate the invoice. The rating engine accepts UDRs from the provider's switches or from another providers system. It then checks the validity of received billing records with respect to format. It matches these billing records to customer in a database. It then provides billing details to the other IT systems used in the billing process. The rating engine is used to collect usage charge information and route rated billing records to specific customer accounts. In order to recognize an account, usually the *Telephone Number* (TN) or other *Network Address* (NA) is used to match the billing record to a specific customer account.

The invoicing business systems will aggregate billing records for a billing cycle. Then, they will calculate recurring charges and produce an invoice to be delivered to the customer. The customer pays a portion of the amount indicated on the invoice. The payment is received and then "posted" into the billing system. The history of payments is recorded in the billing systems as the payments are made.

The *Customer Service Representative* (CSR) in the providers call center can typically go into these billing systems and review a customer's payment history to determine if a payment has been made while interacting with the customer.

How does a provider aggregate billing account information across affiliates when it becomes available? How do ordering systems effectively offer packages and use account and product information to reconcile charges? Well, in order to manage account information across multiple businesses, providers must first establish a common way to model account and product information. They must model the relationships between themselves and their business alliances.

For example, in order to bill correctly, a provider must have a way of grouping billing account information across telecom affiliates. At the same time, they must have a way of grouping product information. After all, they must manage package pricing between all systems involved. See Figure 5.2 towards the end of this chapter. This expresses the relationships. We have an account grouping identified by some identifier. This identifier ties together all of the affiliate billing accounts. This group is generally the rate plan or what is used to identify the rate for products which span across business partners. Then, in turn, each billing account has a recursive relationship (tree structure) of product offerings containing other product offerings or products.

This is how a provider can manage the bundling of products. In addition, this is how it is possible to manage account information which spans across multiple affiliates. It is important to aggregate account information received from each affiliate into one central grouping. If we don't do this, it is almost impossible to manage the necessary relationships. Why do we care about these relationships? The answer is "billing." If we expect to bill correctly and have the invoice reflect accurately the products and features on someone's account, we need to have this information. This will make it possible for the IT systems to manage the billing process. The diagram below shows the typical billing system and it's relationships to the customer and the other business entities.

The important thing to remember is that a "Rate Plan" is the entity which is used to group account and product information across multiple organizations. Affiliate data feeds, and in some cases, real time messaging will allow a telecom organization to receive the appropriate account and product information. They need this information in a timely manner in order to reduce latency in the billing process. In the feed, they will receive the specific billing account information for a particular customer based on a product he or she has recently ordered.

When the information is received, it is important to aggregate account information. It is important to manage the product information received from each affiliate. It is also important to group affiliate account information, associate products to the account, and associate accounts to the provider's billing account information. In other words, affiliate account and product information must be associated with the account the provider has established for the customer. This means the provider not only groups account information across multiple affiliate accounts, but the affiliate product information residing on these accounts as well. This allows a telecom organization to effectively manage product packages.

Providers need to be able to access customer account information. They need to be able to access this information across business partners. They need to be able to modify this account information and establish new accounts for customers. They need to able to associate recent product orders to an account and have this information reflected in the business partner's billing system.

They need to know, that if a product is removed from a customers account, this may disqualify them for a package. This disqualification constitutes a new price. This new pricing needs to be reflected in the appropriate billing system or systems and on the customers account in order to bill the customer correctly.

Why is this really important? Not only for the obvious reason of correct billing. Package offerings are important in general to a telecom organization for customer retention. They want to offer discounts for customers to buy multiple products. Though these products reside across business partners, the provider makes this appear seamless with respect to billing and support. This is a convenience to the customer. It makes them less likely to leave for another carrier.

Interfacing with Affiliate Billing Systems

Providers generally have multiple "Sales Channels." A sales channel is essentially a doorway into the provider's organization. It is used to order products and services. This doorway can be a call center, the *World Wide Web* (WWW), an IVR, PDA, etc. When the customer orders a product or service through their provider, they must go through one of these sales channels. They may call the provider's call center to request the product, order the product over the web, or maybe, they call an automated voice system. Regardless, the customer must first order the product; the product or service must then be provisioned in the IT ordering and provisioning systems in order to have the service available.

Provisioning means the service becomes enabled on the provider's network. It is similar to a light switch in the sense that it turns a service on and off. When the provider provisions the service, they are turning the switch on. The provider is generally making the service available to the customer starting on a specified date. This may or may not be for a specified duration. Then, the customer must be billed for the service. This is done by adding the product or service information to the customer's billing account along with any corresponding rate plans.

The *Customer Service Representatives* (CSRs) work in a telecom call center. They need to interact with the billing and ordering IT systems on a regular basis. This is typically done through the use of desktop or internet browser based software applications. This software interfaces with back end IT systems to process orders and billing requests.

When the order involves ordering products through an affiliate associated with the provider, the service order must go through the telecom providers ordering systems to record the transaction. However, ultimately, the order will go through the affiliates ordering system to provision the service.

The business partner's ordering system will provision the service. Then, it will update their internal billing system with the provisioned product information. The affiliate IT billing system will create or maintain the account information needed to establish the correct billing process. This is to assure the telecom provider is billed correctly for the service. The information is then fed back to the telecom providers billing system. The provider uses the account information from the affiliate to invoice the customer appropriately for services ordered.

The invoicing portion of the billing process generally involves gathering the data from where it is stored. This typically is in a database where collection tables exist. The database holds tables which contain the collected billable charge information. Then, the information related to adjustments, non-usage charges, tax and fees, and usage charges are taken from these collection tables. They are put into a format which an "Invoice Generator" (billing software to generate the phone bill) can understand. The invoice generator makes any adjustments, calculates any volume discounts, and then applies them to recurring and non-recurring charges. Tax and fee information is then added to the invoice. This information is based on the resultant calculations derived from product information. The product information exists as a catalog of definitions. This "Product Catalog" represents all of the product definitions existing in the company, including affiliate products.

The "Invoice Generator" processes the information and creates the phone bill. It is printed and delivered to the customer; the customer then makes the payment. Once the provider receives it, it is "posted" to the customer's account. The process is continually repeated on a regular billing interval commonly referred to as the "Billing Cycle."

Telecom billing systems are generally quite complex. They tie into ordering and affiliate billing systems for reconciliation. They also access other IT systems in the telecom enterprise to sync up information. As a customer makes calls on the provider's network, the network creates *Usage Detail Records* (UDR) which record the activity. The UDR not only comes from the consumer connecting with one of the telecom provider's services, but also from other affiliates. For example, long distance or wireless providers. These UDRs are rated using the billing systems information, based on the customer's rate plan, associated with the particular service. Then, the information is placed into a staging area. It remains there until the end of the billing cycle when it will be placed onto the customer's phone bill. Of course, first it is necessary to aggregate data from all IT system sources before the billable charges can be properly invoiced. Sometimes, the billing information comes from "Clearinghouses."

A clearinghouse is a third party organization which audits or acts as a third party to validate telephone bills for service providers. Especially, when international billing and currency exchange issues are encountered as part of the billing process. They also provide a variety of other functions for service providers. These include reformatting records into proprietary formats and validation of charges. They are the middleman, when providers have billing agreements between one another. Many times, they are involved in receiving billing records from one provider. They will submit them to another provider, after a careful auditing process. This allows the provider to update their billing records correctly.

Billable Events

Billable events represent events which are internal or external to the telecom billing system. These are events which affect the billable amount. These may be an adjustment to the bill from another IT system, or an order for products or services. "Billable Events" are, essentially, any events which a third party may invoke on a telecom billing system. This includes updates, usage events, recurring events from clearing houses and affiliates related to billing. This occurs for prepaid types of services.

Let's say, as a consumer of telecom services, I need to allocate 100 minutes to my son Dalton's Virgin Mobile wireless plan. In order to process this type of rate plan, the provider needs the ability to create an event on the billing system in "real-time" or at least "near real-time."

This event needs to be triggered to submit a service order and allocate the minutes in a short period of time. The difficulty here can sometimes be authorization and authentication of the user. The user must be authenticated with the IT systems, for a small period of time, while the user is authorized to add minutes. The request for additional minutes must be "posted" to the account (added to the account and told to process). Then, the "Rating System," which is processing the usage information, must be notified. The user must be logged off after the minutes are provisioned.

When using services which operate across multiple providers, many times, revenue for those services must be shared. This results in wholesale agreements being made between providers and their affiliations. Let's say I order a satellite with my phone service. The provider needs the satellite provider to offer the product at a discounted wholesale rate. Then, the provider can, in turn, resell the affiliate product or service and make a profit. This is a common occurrence. In telecom when retail services are being ordered, this means services are sold directly to the consumer. When wholesale services are being ordered, this means services are being ordered through a third party affiliation.

"Billable Events" are many times stored in what is referred to as a set of collection tables in a IT database or a "Data Collector." The data collector transfers the information to the billing system when it requests the information. This can be a push or pull operation in the sense that the billing system can poll the collection tables on a specified interval. Then the billing system can retrieve the information from the tables. The information can also be pushed into to the billing system from the process or system which loads the tables. It is also possible to trigger events, which perform actions (common in relational databases), based on different data in the collection tables.

We've discussed UDR information but what about CDR? CDR refers to *Call Detail Records (CDR)*. This includes information other than how long you were on the phone or affiliate service and at what rate. Instead, the CDR contains information like who originated the call and who received it? Where the call came from? Where the destination was? What time of day was it? Etc. In other words, CDRs present information to the billing system to process for the invoice. This is beyond what the UDR processes. The UDR is limited to network specific elements, network elements like usage and rate. The CDR has more information which goes on the detailed portion of the invoice relating to call detail.

Customer Accounts

Customer accounts represent the financial records which exist in the provider's IT systems. These are comprised of attributes for the customer, such as billing address, service address, credit score, social security number, and the products which the customer is being billed for. Billing systems have software which receives the UDRs. They use the UDRs to rate service usage information through a rating engine. They associate the rate amount with a particular usage record. This is based on the customer's rate plan. The rate plan is listed on the customer's account and is, typically, associated with a product or package.

The account may also have recurring billable charges which refer to a monthly (or some other interval) fee, which is for the most part fixed. The rate information, recurring fee information, tax and fee information, and all of the UDR information from affiliates are aggregated. It is then used by the billing system to process the customer invoice, commonly known as the phone bill.

The aggregation of customer account information can be difficult due to the time lag between billing systems. Since this billing systems reside across business partners. When I say difficult, I mean it can be difficult to get the billing cycles aligned, in order to provide the charges a customer expects on a given month.

Billing cycles represent a grouping of billing operations for a subset of the provider's customer base. This breaks up the processing involved in the invoicing operations to be segmented to provide the best benefit for the consumer and the provider. This adds to the complexities associated with billing synchronization between telecom affiliates.

Sometimes mediation devices are used to aggregate information from different billing and ordering systems.

Mediation refers to the collection and aggregation of information across systems. Think of mediation as a collector and a router. It manages workflow and normalizes information across backend systems. Let's say, a provider needs to get information from one telecom IT system and one affiliate IT system. They need a device or piece of software to mediate and collect the information from these systems. Then it must route the information to the appropriate IT system for processing a report.

Customer account information is used to generate the invoice. Invoices contain the quantity of different services, the description of those services, and the amount due. They also include tax and fee information, detailed record charges for usage information, as well as recurring and one time setup charges. It, usually, provides call origination, call destination and detailed rate plan information. If a rate plan is involved, usually the rate plan, which may operate across telecom affiliate services, is listed.

The detail is typically provided farther down in the bill. The detail may or may not be separated by the affiliate itself. At the very least, it is usually described in the description field. It may or may not indicate what products are associated with the rate plan. This can make it confusing for the consumer. When this happens, the consumer has a difficult time determining what they are paying for. Many times, the full price is showed next to the detailed item. The discount is applied somewhere else on the bill, in accordance with the rate plan.

Account Management (AM) not only refers to managing the phone bill. It also involves managing "Trouble Tickets" or the creation of trouble tickets. This maintains a record of what service was provided to the customer and if the problem is recurring. It allows management to determine problems in the field, and allows them to make decisions regarding the best way to address those problems.

Payment Management

Payment Management (PM) refers to how payments are retrieved through a collections group or department. Some providers will go to the extreme of routing calls to a third party collection agency, if the customer doesn't pay their phone bill. The customer will typically be asked to pay their phone bill either partially, or in full, through a variety of payment options. If the customer does not pay within a given duration, their service is usually terminated. Historical payment information is readily available when managing customer payments. We all try to make our payments on time. If we don't, we can expect a "Collections Agent" to contact us regarding payment of our debt. They will generally give the customer many different options to pay. They may offer to accept partial payment. They may indicate that telecom services will be terminated. This becomes common if the bill is not paid in a timely manner.

Credit Classification

The "Credit Risk or Classification" is an important concept. This refers to what "type" of customer are you to your provider. For example, the provider may rate you based on different discrete labels. The labels may range from a superb customer to a poor customer. These designations are generally based on many factors. Factors, such as, do you pay your bills on time? How long have you been with the provider? You're not a credit risk, etc. This is many times referred to as "Scoring" the customer.

Your credit risk many times ties into the types of products your provider may offer you, and at what price. The concept of *Offer Management* (OM) is an important concept. This refers to the situation where certain offers are provided to you. They are based on the type of customer you are. For example, you may qualify for a discounted rate on certain packages if you are a superb customer and always pay your bills on time.

Services

A telecom service is a provisionable or orderable item. It is usually provided by a telecom service provider or one of its affiliates. These orderable services are offered to the consumer at a specified rate. This many times refers to a network type of service. A service can be an individual residence phone line or a wireless service, for example. The service is identified on the customer's billing account usually by a *Universal Service Order Code* (USOC) or some other product identifier. The service is enabled by the telecom network. It may or may not have usage information associated with it. A service can be a telephone line or a video service. It can be anything that must be physically switched on or enabled for the consumer to begin receiving the service.

Products

Products are not provisionalble. They are non-telecom network related by nature. They are not a network serviceable item which can be ordered from a telecom provider or one of its affiliates. It can be offered to the consumer at a specified rate. This, many times, refers to a phone, modem, satellite antenna, telephone directory, etc. Products generally do not reflect any usage charges (time sensitive charges). They are typically one time charges or even recurring charges in some instances.

Think about a "phone" as a product or a telephone directory. Think of a satellite receiver as a product. Products, like services, are generally referenced on the customers billing account. They will be identified as a *Universal Service Order Code* or USOC. This USOC identifier is common in the world of telecom. However, in non-telecom industries, the identifier can be any value which uniquely identifies the product. In order to synchronize product rate and definition information across telecom billing and ordering systems, it is prudent for the provider to assure the information is centrally located. It should be easily accessible from multiple IT systems within the enterprise.

Not only do product definitions need to be maintained, but also the feature information associated with a particular product. For example, each product has a set of attributes associated with it. These attributes can be a name, locations where the product is available, the credit class of customers which can order the product, etc. Each product can have not only attributes, but it can have anywhere from zero to many features associated with it. For example, if I have an individual residence line and I need to add call forwarding to that line, this would be a feature of the individual residence line. Now, call forwarding in itself, may have attributes associated with it that make it unique. For example, how many rings are allowed before the call is forwarded to another destination.

Not only can a feature associated with a product affect its provisioning, but it can also affect its rate. If I, as a consumer of telecom services, select different options when ordering the product, this may force some implementation complexities. For example, as a result of my feature selections, the provider may be forced to place the product on a different network element or telephone switch. This may be due to lack of capabilities from the network element. This implementation change may be susceptible to a differentiation in rate. So, my bill is increased as a result of the feature selection. This is primarily due to the change in switch configuration and the resultant complexity.

Packages and Promotions

Packages are a grouping of products. This is not limited to a telecom providers products, but cross affiliate products as well. Packages, themselves, are many times, a subset of a rate plan. The rate plan typically groups business partner accounts and the products associated with them. This is for the purposes of billing. Packages refer to a grouping of products which are then tied to a customer's account. Generally a new *Universal Service Order Code* (USOC) is created or another packaging identifier.

The customer may order one or more packages which contain products which belong to one of the provider's business partners. Usually, the telecom provider's ordering system, which is used to order and provision the package, will update the telecom provider's IT billing systems. This includes cross affiliate product and package information. The USOC on the account will refer to the set of products required to make up the package. In other words, it may reference another USOC or product identifier.

A promotion on the other hand is considered any combination of products or packages. This is, in addition, to other types of offerings. For example, as a consumer of telecom services, I get a ticket to my favorite movie if I buy a product or service. This is a promotion. I may buy one package, one ala carte service, and an extra directory listing. If I do this, I get the ticket to the movies for free.

Rate Plan

A "Rate Plan" is sometimes referred to as a "Pricing Plan." This refers to how the customer is billed for products and services. It, many times, includes business partner products. The rate plan, for example, may indicate a

$29.99 dollar a month recurring charge for wireless service. In addition, the customer will get 1000 free minutes during off peak hours. The rate plan may also include a high speed data line which operates at 1.56MBPS. The rate plan may include a land-line. All of these services may be provided under "one" rate plan. This means, in order to bill the customer effectively, the account information must be synchronized. It exists across business partner IT systems between the telecom provider and the affiliate IT billing systems. This information must be collected and grouped under one rate plan to invoice the customer correctly. It gives the provider the ability to place all business partner products on the bill. This includes pricing information such as tax and fees and rates.

Product Management and Marketing

Usually, a telecom provider has a group of individuals responsible for marketing products.

Sometimes, this department is referred to as "Product Marketing" or "Product Management." These individuals will determine how the product is defined, its pricing plan, and any promotional offers associated with the product. They will also determine what rates are associated with this product or package, etc. These individuals then use software to update IT systems with this information. This allows sales channels to utilize this information when ordering products for the customer.

It is important that the product information is accessible by all sales channels in a consistent fashion. Providers don't want sales channels making offers to consumers inconsistently. As a provider, I don't want my self service web sales channel, which allows people to order telecom products and services over the internet, to offer different products than what exists in the call center. At least, a provider does not want to do this, unless the product is sales channel specific. For example, the provider may have a product which they are using to eliminate calls coming into a call center. In this case, they may want to offer a discounted rate if the product is ordered over the web. This allows them to direct traffic to a particular sales channel and minimize costs associated with running the call center.

Account Management

"Account Management" refers to the ability to manage billing account information. This includes customer attribute information such as name, tax identification, service address, and billing address. It also includes the products the customer has existing on the account. This information is usually represented as the *Customer Service Record* (CSR).

The CSR is an old Telco record format indicating how customer accounts should look. It defines a consistent format for providers to use. Affiliates which offer services beyond basic Telco type of services, may utilize different formatting conventions, other than the CSR.

Management of account information is not an easy task. For example, the customer may have an account which references other accounts. Many times, the account information incorporates a cross referencing mechanism which allows one account to associate itself to another account. The account can refer to another one of the provider's accounts or an affiliate account. The other account identifier exists as a listing on the account as part of this association. Maybe on a customer's account there are Telco products, but also services and products which belong to the provider's business partner as well.

Other than business partner account identifiers, generally, some other type of identifier, which represents affiliate products, will exist on the account. This can be a *Universal Service Order Code* (USOC) or some other identifier. The identifier can then be used to access the affiliate IT systems and retrieve the full record of the account the customer may have with the affiliate.

Flex Bundle

Packages typically are made up of particular products. However, sometimes they can be made up of a selection of products. In other words, the customer has a choice to choose a certain number of products which comprise the final package. This is referred to as a "Flex Bundle."

An example of a flex bundle is when a provider offers product a, b, c, or d at $29.95. Then allows a customer to choose any number of products to represent a, b, c, or d. It provides a choice but associates the package to a single price. For example, maybe a package consists of a phone land-line a wireless phone w/service, a satellite TV dish subscription and long distance. The provider may offer this entire package for $49.99. The provider wants to provide good customer service and allow some flexibility in their packaging of products. As a result, they offer you the choice to substitute one of the products with any other product they offer of equal or lesser value than the lowest priced product. The customer then chooses to replace the satellite TV service with high speed DSL instead. This situation is a good example of a provider offering a flex package. The products which make up the package are not fixed in this case and substitutions can be performed.

Bolt On

A Bolt-On is a term used when adding a product which is not included in a package offering onto the customer's account. For example, let's say I have the Top 100 package with the DISH Network. It includes 100 channels of the most popular programming. However, against my wife's advice, I decide to order the Playboy channel in addition to the Top 100. In this situation I'm requesting a Bolt-On. This means I'm paying full price for a service on top of the package which is receiving a discount.

Vertical Service

Vertical Service is a service which sits on top of a product. It is many times a dependency from one product to another or sometimes a feature dependency of a product. A typical telecom product can have multiple vertical services associated with it. For example, let's say I have an individual residence line. The residence line has call waiting, call forwarding, call return, etc. These would be considered vertical services or features of the residence line.

Tax and Fee Management

Tax and fee management is the ability to determine the applicable taxes and fees associated with a product based on the service location. This information is used by the rating and sometimes the invoicing engine to determine how to calculate Tax and Fee information on the customer's telephone bill. This includes fees for hardware usage, state, federal and local taxes and other fees such as cross intra-carrier types of charges. This includes fees like roaming charges on a wireless network for operating on different hardware or switches during the duration of a telephone call. Tax and fees are many times administered by federal, state or local city taxing bodies.

They may be related to telephone surcharges, state or federal taxes, or user fees. They are many times tied to location or hardware usage. They may also be tied in the inter-LATA or intra-LATA types of usage services. When a phone call is routed through a phone system, it will be routed through various locations and incur different surcharges and taxes along the way. These taxes and fees are usually calculated by the billing system invoicing engine software which generates the customer's phone bill. The fees are generally added to the bottom of the customer's bill.

Party Management

Party management doesn't refer to how to setup and coordinate a party. It refers to party in a general sense. An organization can be a party; a person can be a type of party. A party represents an actor in a business process. The actor can be a customer or a telecom employee. It can be an operational support system. Essentially a party can have roles and relationships to other parties thereby allowing for various relationships between business processes, IT systems, and people. The concept of a party to represent an individual or an organization is used heavily in IT systems. Customer information many times must have different relationships. These relationships are somewhat dynamic to different business entities.

Package or Bundle Management

The term package or bundle management generally refers to the ability to manage what products exist or don't exist in a particular package and the pricing rules associated with those packages. Telecom ordering systems must be able to determine whether the customer qualifies for a discount due to a package they have ordered. For example, if I order an individual residence line and a wireless phone I may qualify for a 10% discount if I buy these products together as a package vs. individually. If the customer comes back and removes a product from their previous package order, the ordering systems or business processes should be able to determine that the customer no longer qualifies for a discount and now must pay full price for the products or services.

Location Management

Products and services are many times offered only at particular locations. For example, newer high speed data technologies such as DSL are only available in locations where the telecom network can support it. This is a situation where the location or service address determines if the customer can order the service. Many times, when dealing with multiple affiliations, a telecom organization must provide location information in different formats. Each IT system supporting the product availability at a service address generally processes location information in a different way.

If a telecom service provider is offering packages of products which cross affiliate boundaries, then the availability of products within the package must be reviewed to assure the product can exist at a service address. Maybe, I want to order a high speed data line into my home. I see a telecom provider is offering a package for $59.00 a month which includes a high speed data line in addition to residential phone service. I then decide to order the service. I call the call center for the telecom service provider and ask to order the "Data Package." The "Data Package" is then added into the ordering system the Customer Service Representative (CSR) is using to order products for me.

Before adding the package to my billing account, they first much check to see if each product in the "Data Package" is available at my service address. If it is not, the CSR informs me that the service is currently unavailable or tells me when it will be available. I'm unable to order the "Data Package" but I can still order any services in the package which is available at my service address.

Many times location information is broken down by different regions. For example, all products in a certain state are offered. Maybe it is broken up by east, west, north, or south. There are a variety of ways to tie a service address to a region. In the case of phone service, generally the actual switch which is providing the service and then routing the call determines the capabilities. If the switch which provides service at my address supports "call forwarding" for example, then I can order call forwarding at my service address. Availability is not always determined by product, sometimes it is by features the product supports.

Consumer vs. Complex Ordering

Consumer ordering is one aspect of ordering within a telecom organization. The other is "complex ordering." Complex ordering involves ordering products or services for large businesses. It may include specialized operating equipment to transport communication signals over the phone line at a high rate for example. It may support direct dedicated phone lines to support reliable communication. Maybe a key system is installed at the customer's location and the system needs to be designed and configured.

There are a variety of other important business reasons for supporting complex ordering. Complex ordering usually requires business processes and workflow for those processes to be successful. In the telecom organization, there are many departments which support different aspects of the ordering process. The workflow routes requests between the different workgroups in order to effectively process orders. Complex orders don't have as much automation as consumer orders due to the specialized requirements. The solution is typically unique to a particular business so automation cannot totally encompass the solution though it can automate individual business processes to come up with the finished solution.

Consumer ordering generally involves residential retail or wholesale services and even small business in some cases. It can be considered more of a canned solution. The requests for provisioning of services on the telecom network generally can be completely automated. For example, if new phone service is requested and the customer is new to the provider, the customer's account information must be setup. This includes information like tax identification, service address, billing address, etc. This operation is commonly referred to as a "new connect." If the customer requests changes to their account with respect to the services they are currently being provided, the CSR looks up the account information. The information returned to the ordering system is their account information. The CSR then makes changes to the account and submits the order to IT systems to process the order and disable or enable services on the telecom network.

Offer Management

Managing offers involves determining what offers and promotions are available for a particular customer. This may be based on the customer's credit rating or other factors. For example, I've been with my provider for 5 years and I have always consistently paid my bills on time. This fact alone may incline the provider to offer me a discounted rate on special promotions to keep me as a loyal customer. Providers generally use some type of rating system which addresses the "type" of customer you are. This may be a rating system from 1-4 or 1 -10 with varying levels of rules associated with each customer rating.

Let's say if my customer rating corresponds to a "1." According to the provider this means I've been with the company for more than 12 months, and I pay my bills on time. The provider's ordering system generally will provide a different selection of promotional offers just for the "type " of customer I am since an identifier on my account indicates that I'm a "1." The other factor that comes into play is availability. I can only be offered products and packages which are available at my service address. In the case of mobile products such as cellular phones, the location can be based on region. The ordering system will generally filter out any promotions for which I'm not qualified. A telecom provider's marketing department can come up with many other factors as well regarding what promotional offers to provide to a particular customer.

Adjustments

This refers to how payments are managed through the provider's call center. In the call center, when a customer calls in and there is an error on the customer's phone bill, the representative has a mechanized way of handling payment adjustments. They generally have a way of looking up the customer's payment history and determining if an error has occurred. If it has, the adjustment is submitted to the billing system. Billing adjustments imply a credit or debit is needed depending on circumstances related to the bill. For example, you may call a CSR in the call center and indicate you have an error on your bill. After doing some research, the CSR may make an adjustment to your account and the adjustment will be reflected on your next billing cycle. Payment and adjustments are usually managed by a Payments and Adjustments IT system which provides information with respect to your payment history. The system generally will generate a billing event to update what you need to pay at your next billing cycle. This information will, in turn, be placed on your next invoice.

Usage Management

Usage Management is the collection of charges off a telephone switch or third party affiliate product which is time based. For example, I get 100 long distance minutes free each month. The billing system needs a way to record this information. Usually, this is done through the use of a real- time rating engine tied into the customer's account. The information is fed in real-time indicating how much of a given service the customer is using. Let's say I talk for 10 minutes. At regular intervals the telecom network where the service is operating must record the duration of the call. In the case of prepaid phone cards, the maximum time interval information must be distributed to the telecommunications network so the call or service can be terminated when you reach your maximum usage value.

Billing Cycle

Your Billing Cycle (reflected on your bill) indicates how often you are billed for telecom services. It is a recurring time period for example from the 1st to the 31st of each month or from the 15th to the 15th of each month. It is a cycle of time which all charges are recorded and at the end an invoice is generated and sent to you, the customer. Why do provider's use billing cycles? Think about the amount of processing which must be performed in order to generate invoices to so many people. Using more than one billing cycle allows a provider to divide up this processing; increasing the ability of their IT Systems to support a greater number of customers.

Rate Plan

As a consumer, probably the most important decision you have to make when requesting telecom services is deciding on a rate plan. This is the plan that determines pricing per usage, recurring charges, service charges which are associated with a group of products. Many times this can be the basis for a package. I may sign up for the 1200 minute a month rate plan where I have to pay 29.99 a month recurring charge for the phone and 1200 minutes a month and only incur usage charges when I exceed the 1200 minutes. This type of arrangement is referred to as a rate or pricing plan and is many times comprised of many pricing elements.

Invoicing Engine

The invoicing engine uses a definition of billing account information to determine how a bill is defined, formatted, and distributed. Many times, this involves synchronizing billing information across telecom business partners in order to allow one phone bill to be distributed to the consumer representing charges from different businesses. Let's say I order a high speed data telephone line and a satellite dish service for receiving high definition television. I would like to have the phone line and the video service on the same bill from the same provider where I ordered the bundled service. I don't want to receive one bill from the telecom provider and another bill from the telecom affiliate offering the satellite. After all, ideally, I just want one bill reflecting both services. This sounds simple, but many times it is not. Different businesses operate in different ways utilizing different terminology, different business processes, and IT technologies to process billing information. They may have different billing rules, fees, taxes, and most importantly billing cycles. The billing cycle which represents how often you receive your invoice for services, is many times different between the telecom service provider and their affiliates. The satellite provider may only bill once a month for services where a telecom provider may bill every two weeks. This makes alignment of the amount charged on the phone bill for a given month difficult. Providers want the phone bill to represent what the consumer expects to see in a given month. The consumer doesn't want to see partial billed items due to dependencies across billing cycles reflected on their bill. Formatting for your bill or your bill definition is tied to a grouping of accounts by a pricing plan which encapsulates these accounts. These accounts may even span across affiliates within your providers organization. The difficult billing problem for providers is syncing up billing cycles across affiliates. For example, if you order an affiliate product from your provider. You want satellite TV for your home and your provider just happens to have a great deal on telephone service when you also purchase a satellite TV bundle or package with 100 channels. When the provider attempts to provision the product or service, they must send a service order request to the affiliates IT system. This is usually done through some type of *Business to Business* (B2B) interface.

The affiliate must then provision the requested service and update their internal billing system and indicate what provider was the originator of the order. Then, the service provider which originally ordered the product must be updated with the billing information on a recurring basis whether the information be usage or recurring related. The reason is not only for affiliate to provider billing, but for the sole purpose of providing the ability to provide one telephone bill which represents all services. You wouldn't want to receive bills from your provider as well as the affiliate that offered video services, so business processes must be put in place to reconcile these billing operations between the two companies.

Your bill is usually generated based on your CSR or *Customer Service Record*. This record is stored in backed *Operational Support Systems* (OSS) which manage your account and customer information. Each section in the CSR will match up with a similar section on your invoice. The products and services and the line-item charges on your bill will be represented. This will include recurring, usage, and one time charges as well as taxes and fees which may exist.

Rating Engine

Rates and discount business rules can easily become complex. The provider may charge the consumer for services or products individually. But if the consumer orders a "package" then they may automatically qualify for a promotion and get a discount. For example, if I order the "family plan" which includes product a, b, c, and d, I would get a substantial discount by ordering them as a "package" or grouping vs. purchasing them individually. A rating engine manages rates associated with products and the features associated with those products. The rating engine must operate near real- time in many cases, especially in the case of prepaid telecom cards or other services. The rating engine must be able to rate products, determine applicable tax and fee consequences, and work across telecom affiliates, in some cases, in order to effectively rate packaged services. The rating engine is a common point in the telecom business architecture which allows ordering software across sales channels within a telecom organization to rate products in a common way.

After all, a provider must be consistent in how products are rated whether the order is placed from the internet or via a service representative in the provider's call center. One concept, which is important, is the "prepaid" vs. "postpaid" concept which is new to many providers. For example, if I need to download ring tones to my cellular phone, I need to pay upfront. Maybe I want to prepay for minutes as well. These are examples of prepaid charges. Post paid charges reflect mostly usage. You don't pay up front instead you pay as you go. One important calculation is how many minutes do I have as I'm speaking at a given time, and what is the max amount. The service needs to be disconnected when the max usage is utilized. In order to disconnect the service and record usage, the rating engine needs to be in close contact with the switch as well as the rest of the billing system. The billing process involved with rating is typically comprised of collection and formatting of usage information, transferring the records to the rating engine which will rate each record received to the billing system.

Billing Reconciliation

Synchronization and reconciliation is important when considering how to align billing information which resides in telecom affiliations operating against different billing cycles. It is common for telecom businesses to operate on different billing cycles. In other words, at what recurring interval do I generate an invoice to the customer for services rendered?

The synchronizing of billing information from affiliates to the telecom provider can happen in a number of ways. One common way is the "batch feed." This involves dumping what exists in the affiliate billing system to a "file" similar to what you see on your personal computer. The file is then sent to the telecom provider in an automated fashion using something called a "file transfer protocol." This is similar to how files are transferred from one machine to another machine over the internet. Once the files are received at the provider's location, they are loaded into the provider's billing systems. Many times the information is first staged in a database and then the billing system can retrieve the information it needs from the staging location.

Another, more advanced way, of receiving information from an affiliate is through messaging technology. This technology can synchronize up information between IT systems in real-time. In other words, when the telecom service is initially ordered, the billing transaction is reconciled automatically. To review an example of this, let's say you order a high speed data phone line and satellite television as before. When the order was placed through a representative in the call center, the rep would enter in the order into software which would process the request. It would send a message to the satellite affiliate indicating the order had taken place. The satellite provider would then respond immediately to the request so that the provider could synchronize up the telephone bill. How does this messaging work? IT messaging works similar to how "chat" software works on the internet. When you send text messages to someone called "Cyberdog" and he responds, this is a form of messaging. IT messaging between the telecom provider and the affiliate happens in a similar manner. The only difference is that the message is guaranteed to be delivered. In other words, the message will always reach its destination by using safe guarding techniques. For example, if the other system is not available, then the sending system will continue to send the message until the system is available.

When different technologies exist between providers, they must find a way of effectively communicating messages between one another using an IT messaging system. In order to isolate these different technologies, usually "web services" which define a wrapper around different computer languages and components, is used to standardize how the systems will communicate.

Web Services essentially wrap computer languages in something called XML (eXtensible Markup Language). The XML can be interpreted by both systems easily in their respective computer languages without any problems. This has become a common way of communicating between a telecom provider and their affiliates.

Probably from a billing perspective, the most difficult aspects of synchronizing up-billing information from the affiliates to the telecom provider, is how to interpret billing account information from other systems which may interpret account information in a different way using different terminologies and sometimes different business entity relationships internal to the billing system. This can be difficult and involves mapping one system to another with respect to information. If I call a credit score a "credit score" in one IT system and a "credit class" in another mapping, I need a way to say that "credit score = credit class." This is done through the process of mapping. When I map something, I'm saying that I will give a way for another IT system to interpret what I mean. This can be done through technologies such as XML, a database which stores information and its relationships or any basic computer "file" format. Once I define the mapping, I need a way to represent the information to the billing system in common fashion. I don't want the billing system to need to interpret information from different IT systems. I want the billing system to expect and interpret information one way and one way only.

So I place the mapped information into a common file format and I model or define the relationships in a common way no matter where the data came from. For example, if XYZ company defines billing information in one way and ABC company does it in another, I map the information to a common set of terms, and then model (express) the relationships so that they exist in common way that the billing system understands. A common way to think of a generic representation of a customer is through the concept of a "party." A "party" can be an organization or a person which can have different types of relationships to different IT systems, other parties, and can have different roles with respect to each. So as a "party" I can have an employee relationship to an organization.

I can also have a relationship to a billing account, an address, a spousal relationship to another party, or any other type of relationship. Maybe, I have a relationship to a grouping of accounts. Maybe in this grouping of accounts I have a relationship to be used to define the types of discounts I receive. Or maybe this information is used to synchronize up billing information with third parties.

You can think of it this way. I as a consumer of telecommunication services exist as a party. As a party I have an "account" relationship to one or more billing accounts. Each of these accounts can be a different affiliate. As a party, I have a billing address, a service address and attributes associated with my account and myself as a party. All of these relationships group the account information together and identify them as having a relationship to myself. But how does the account information group itself to a telephone bill. After all, I may have different phone bills coming to my house for different sets of services I've ordered. In this case, I need a way to group account information. What is the best way to do this? Well, the best way to think of grouping account information from a billing perspective is through the concept of a rate plan commonly referred to as a "Pricing Plan."

A "Pricing Plan" or Rate Plan can group the account information together and the products tied to each account to form a definition of information and pricing needed to create an invoice for the customer.

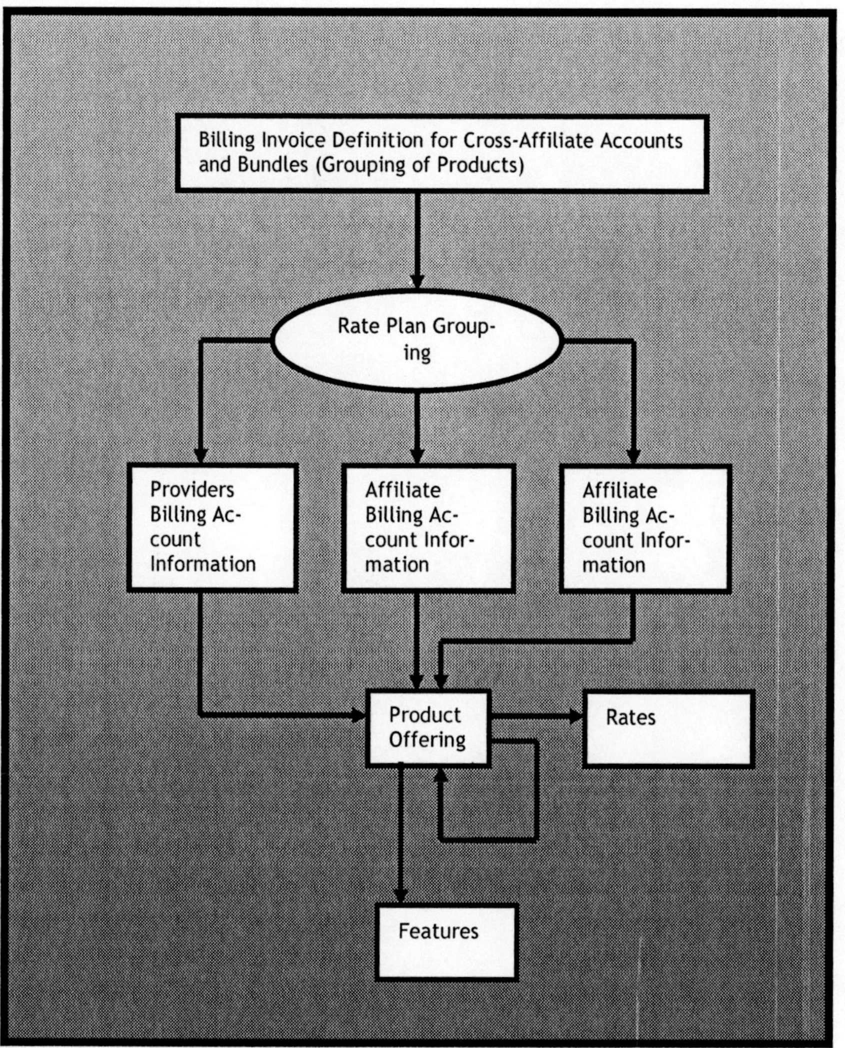

Figure 5.3 Rating plan relationships across business partners

Chapter

6

Customer Care

Intro to Customer Care

Customer Care can be correctly defined as an approach to acquire new customers, provide customer satisfaction while at the same time building customer loyalty. A *Customer Service Representative* (CSR) will work in a call center managed by the provider. They will work on the phone with the customer for sales and support operations.

Essentially, the customer care group in a telecom organization is responsible for the provider's relationship with the customer. They perform a variety of customer support functions from ordering and billing support to troubleshooting. This is an extremely important business function. It is important that the provider remain focused on the needs of the customer in order to remain competitive. They are the customer's window into the provider and represent the view the customer has of the company.

Representatives in the call center need to interface with many different IT systems. They need to be able to retrieve and maintain the information necessary to provide good customer service. In order to further manage the customer relationship, providers collect information regarding their customers spending patterns and personal attributes to determine what in the future the customer may order. For example, if the customer has a teenage daughter, then he/she may be prone to order a second phone line. The provider needs to know what to offer and when to make the offer. In addition, CSR's can use account, historical, and other customer data to determine what to offer the customer when they are on the phone.

This is referred to as "up-selling" the customer. There is a high probability that the customer will order more services from the provider if the information that the provider has is accurate. This involves decision making at the lowest level. It involves using an IT system to determine what is reasonable for the CSR to offer this particular customer.

One important aspect of a CSR's job is the ability to retrieve company information at the click of a button. Since their time on the phone with the customer is limited, access to information related to the business is crucial.

Knowledge Management

The distribution and access of knowledge in a telecom billing or customer care organization is referred to as *Knowledge Management* (KM). For a telecom billing organization this is extremely important. This is also very important for a CSR working in a call center. The CSR needs to be able to determine what products and packages are available in different locations and at what offer rate. For example, if the customer is moving from one location to another (referred to as an F & T in a Telco) then the CSR needs to determine if the same products the customer currently has will work at the new location. Information from different areas of the business such as product marketing changes on a regular interval as the provider runs new ads and marketing campaigns evolve.

Usually, the CSR has a repository of knowledge at their finger tips to meet the needs of the customer. It may pertain to DSL modems, phone technical details, specifications, or anything related to a product or service such as pricing information. Commonly, this would be a web based online tool which could be accessed from any CSR desktop. It would be used extensively by representatives in a provider's call center to look up reference information. Knowledge management software is many times a separate process from the ordering process itself, though ideally it is integrated into the ordering application for the CSR. It many times can become part of the preorder or presales business process as well.

Preorder and presales business processes are comprised of business operations which are performed prior to submitting an order. Once these processes are complete and the order is submitted to the OSS, the operational support system will provision the service on the network and update the billing system to invoice the customer.

Business Partner Software

Generally, a business partner for the telecom provider will provide software to order its products. Many times this comes in the form of a web software application. This software gives the provider access to the business partner's *Operational Support Systems* (OSS) to provision and bill for services. As a side note, keep in mind this access is not generally directly to the OSS itself, but typically through some type of software used to interface and manage the sequence of events to these IT systems.

These OSS systems do the "provisioning" or "service enablement" against the affiliate's network infrastructure and update the business partners billing system. From a business partner's perspective, this is a good approach. It allows for them to distribute software to any business partner and give them the ability to provision and bill for their services. From a provider's perspective, this allows them to instantly begin to offer a business partner's services. This is equally beneficial for both parties. From a telecom architecture perspective, there is a disadvantage, however. The CSR for the provider many times will need to go into a separate IT system to order each individual product. This is great when ordering individual products. But consider what happens when a customer orders a package consisting of multiple affiliate products.

In this situation, they would need to go into each IT system individually to support the customer when the package was made up of affiliate products. This can be difficult to manage and is not generally a cost effective approach to customer service. Many times it is the reality however. This type of arrangement for a provider makes it extremely difficult to do what is referred to as "Bundle Management." Managing bundle information involves the ability to manage discounts for packages associated with a customer's account.

Moreover, in order to manage bundle arrangements comprised of a combination of the provider's products and the affiliates, it is necessary to have all product, account, rating, and tax and fee information managed through one central ordering/billing system. By using separate affiliate software applications to interact with a business partner's ordering and billing operations, the information is not integrated to the extent it needs to be for managing packages. At least not to the extent necessary to manage packages within the provider's ordering and billing systems.

Bundle Management

Bundle Management refers to the ability to determine what products make up a package, what discounts are being applied to that grouping of products, and when to remove the discount if a product is removed from the customer's account. This makes managing orders which span multiple affiliates difficult. This problem is not limited to ordering. Payments from a customer, account management, and general billing questions must generally be addressed using a separate software application provided by the affiliate to access a different set of *Operational Support Systems* (OSS). The diagram below shows a typical telecommunication providers support infrastructure. Typically the CSR must access multiple systems from their desktop in order to effectively support the customer.

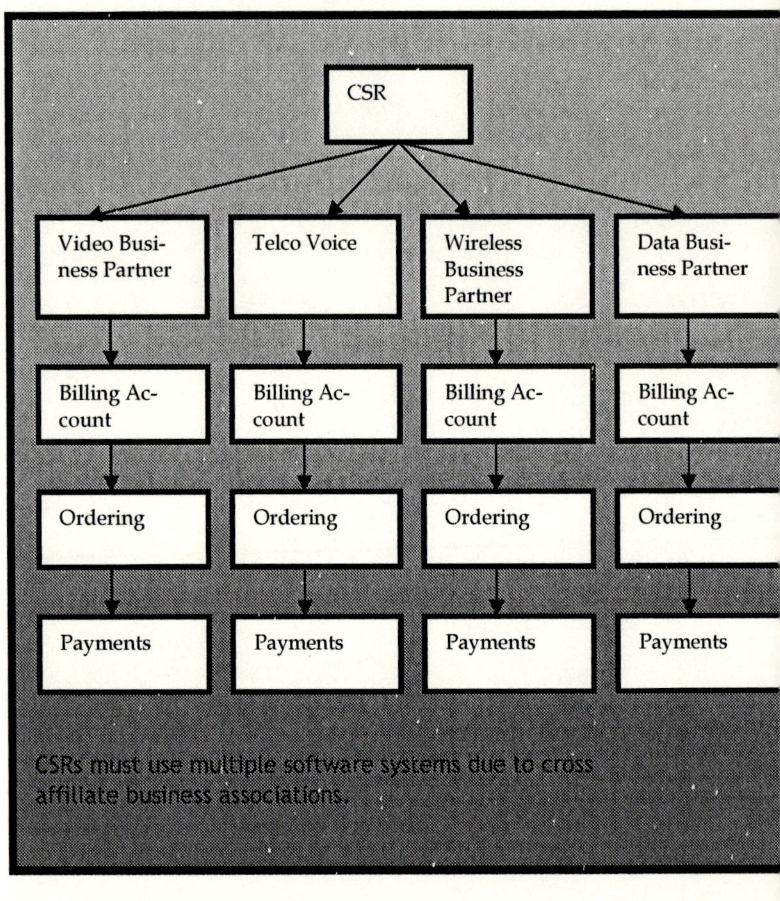

Figure 6.0 Business partner software used in provider call centers

Call Center Software

From a business cost analysis perspective, the more software applications a CSR must operate the more training involved, the more complexity, the longer the call duration will be. While on the phone with the customer, if the CSR must use a separate software application to perform operations for billing, provisioning, payments, and yet another for credit checks, the CSR's ability to provide an adequate quality of service is almost impossible.

A CSR must be able to support people quickly and little training should be required. Many times, when multiple software applications are involved, they have been developed with the ultimate flexibility in mind for their particular business area. Unfortunately, with the fast paced CSR environment, this causes mistakes, increases training time, and may reduce overall customer satisfaction.

In order to understand the complex issues associated with customer management and billing arrangements established for a customer, it is first necessary to review the business relationships between a customer and their provider. A customer or party has a set of established relationships. These relationships relate to the individual's or companies billing account, the products and services they have subscribed with the feature set for each product.

A customer has a set of relationships to records within the provider's *Operational Support Systems* (OSS). They typically have various types of relationships. For example, they may have a spousal relationship to the billing account. They may have a relationship to a rate plan or grouping of accounts which reference different products they have subscribed to. They may have a relationship to an account which the provider's business partner maintains.

A customer care organization incorporates many *Business Support Systems* (BSS) which must exist to assist *Customer Service Representatives* (CSR). A *Business Support System* (BSS) is a business process including business automation software which assists with automation of customer care initiatives.

Customer care organizations generally support multiple sales channels which allow a customer to order products via self service approaches or through the help of a knowledgeable CSR in the call center. This includes payment corrections, payment acceptance, package recommendations, as well as presenting affiliate sales information to the customer to hopefully "up-sell" them to an additional set of products or services offered by the carrier or one of it's affiliates.

Generally, call center software applications can be either web based or desktop based. Many times, these software applications residing on a CSR's desktop access a provider's legacy mainframe applications. This is where much of the provider's business rules exist associated with operations such as billing, credit checks, provisioning etc.

Business Rules

Business rules which reflect product offerings such as rating should be kept in a centralized location which can be accessed by each of the provider's sales channels. The ordering systems used in different sales channels must be able to rate consistently using an interface into the rating engine. It is important to be able to create and easily maintain business rules in a centralized place so that changes can be reflected in each sales channel immediately. Providers also must simplify product rules and offerings to minimize complexity and reduce costs.

Many times, for complex products, where the customer's phone bill may generate calls into the provider's call center, "bill inserts" (messages on the customers invoice) are a way to reduce call volume.

Resource Allocation

In order to remain competitive in an environment with slimmer operating margins than ever before, business priorities of providers are constantly changing. In order to adapt, providers must construct the necessary infrastructure to meet the changing needs of the business.

As establishment of the infrastructure is important so is correctly aligning resources with the appropriate business strategies. Proper allocation of resources is critical to being successful. It is important to establish fixed and dynamic resource allocation schemes to redirect resources as priorities change.

Quality of Service

How do providers supply a good *Quality of Service* (QOS)? In order to provide a high level of service, the CSR must present himself or herself in a friendly manner. They must not transfer calls to others or put the customer on hold for a long period of time, and they must be able to provide full support to the customer during the support call session. The ability for the CSR to effectively use the tools to perform his or her job is crucial. The CSR must be able to access a variety of information in order to meet the needs of the customer.

Standards

When developing a new customer, care system technologies should be chosen that are technology and platform independent to assure flexibility. New software applications should be built using J2EE or .Net framework technologies as a general guideline to assure a flexible customer care architecture. Many providers current standard architecture is J2EE which is a very flexible environment for ordering applications operating across heterogeneous environments.

Though, through the use of architecture, providers always want to adopt a capable solution, they also need one that will have continued ongoing support within the industry. This direction is enforced through the use of standards within the industry such as the OASIS and JSR organizational standards bodies. These bodies govern standardized software interfaces which software vendors will adhere to.

Design patterns are proven reusable techniques for developing reliable telecom software. They are many times documented and expressed with industry based examples. These design patterns are many times implemented over and over again in object oriented frameworks such as J2EE and .Net. These frameworks offer unparalleled support from a variety of vendors and associations which make them ideal candidates for use with customer care software.

In addition to using applications with core industry accepted technologies, CSR's generally utilize top of the line PCs to respond quickly to customer inquiries. They also use Artificial Intelligence software to assist them with "scripting" and provide them with instant solutions regarding how to react in a given situation based on business rules. Scripting involves sentence prompts which are provided to the CSR via software based on responses and needs exhibited by the customer on the phone. All of these basic elements provide a good impression to the customer and improve their overall satisfaction with the provider's services.

Managing Customer Retention

When customers have multiple products with one provider, they tend to stick with this one provider. Billing is a major motivator in this area. If a customer only has one provider which supplies all of their services, then they only need to deal with one provider for billing, support, and ordering. A customer who sticks with one provider due to combined services and billing is referred to as a "sticky" customer. Providers constantly review business processes and determine strategies which can be used to assure a customer stays with the provider. The key process elements are generally as follows:

Product Marketing	*Involves the processes which identify the business rules associated with a product and its configuration. Involved with sales campaigns to market product offers.*
Offer Management	*Determines what offer to present to a customer based on the history and status the customer has with the provider.*
Customer Relationship Management(CRM)	*A customer care business process specifically geared towards customer retention.*
Billing Account Management	*Management of a customers billing account information based on what the services the customer has subscribed to.*
Trend Management	*Customer segmentation analysis and customer historical trend analysis.*
Cross Affiliate Product Management	*Management of product offers and packages across mul-*

	tiple affiliate business entities.

Companies today which offer telecommunication services typically have what is referred to as "stovepipe" business flow in addition to multiple desktop software applications associated to their customer care processes.

Providers must manage packages and promotional offers which include not only their own products but incorporate a business partner's products as well. This concept in and of itself includes addressing complex integration scenarios across affiliate business entities while providing the appropriate reconciliation mechanisms between the provider and the business partner. The provider must sync up the needs of ordering and billing business processes in order to effectively rate, provision, and invoice for the entire list of services in a cross-affiliate package.

Providers must map out the differences between IT Systems across business partners. This is sometimes referred to as "normalization" of information. If the provider's business calls a customer a "party" in one system and another provider calls a customer a "constituent" in another system, both IT systems need a way to effectively converse.

Pricing and Data Normalization

Providers must mediate, normalize and aggregate service order information consistently between the OSS and the sales channel IT systems used for ordering and billing. Why? A provider must manage pricing and billing changes to a customers account when the customer no longer qualifies for a bundle discount? If they receive a discount for example for initially ordering "the family plan" and then they remove a product from this plan, the appropriate business systems need to get updated to reflect the new pricing and product configuration on the customer's account.

It is important to present one bill to the customer which accurately reflects the new rating information for the package. This requires alignment of billing cycle payment information between business partners to have the invoice accurately reflect the correct amounts.

The amount of interactions between different business processes and business partner IT systems is significant to make this happen. Sometimes the systems interface with each other in a "batch" mode where the provider gets a daily feed of the business partner's product rate or account information. Other times, the interactions can be in real or near-real time where a message is sent to the business partner immediately as services are ordered.

Chapter

7

Regulating Authorities

AT & T

In 1885, a new subsidiary company of American Bell was formed in order to expedite the availability of long distances services. This company became known as the *American Telephone and Telegraph Company* (AT&T). The company operated out of the state of New York. As the company grew over the years, fierce competition began emerging in the telecommunications industry. In 1909, AT&T was pushing for "Universal Service" for telephone operations across the United States.

They clearly recognized that the system proposed was a universal, integrated monopoly and would not meet with public approval without some form of public control. So, in order to thwart competition, AT&T embraced state regulation. However, in amidst of their great ability to sell their idea in 1934, the legislature passed the Communications Act, which created a new independent regulatory agency, the *Federal Communications Commis-*

sion (FCC). The FCC quickly initiated the first comprehensive government investigation of the telephone industry.

Today the telecommunications industry is heavily regulated by external bodies such as the *Federal Communications Commission* (FCC) and the state *Public Utility Commissions* (PUCs). These governing bodies continue to determine the best way to assure competition is fair and that growth of newer technologies is not thwarted.

Regional Bell Operating Companies

In the past, and to some extent today, the telecommunications industry and its infrastructure is considered a monopoly. It was determined that telecommunication service providers needed to be tightly regulated to prevent unnecessary price gouging and to assure an adequate level of service for the consumer. Today, it is almost impossible to find a company that is not dependant on reliable phone service.

The terms *Local Exchange Company* (LEC) and *Regional Bell Operating Companies* (RBOC) are "somewhat" interchangeable. Originally, most of the telecommunications operations, whether they were local or long distance services, were handled by *American Telegraph and Telephone* (AT & T). However, in 1984 the FCC split up the company into 7 different RBOCs and only allowed the long distance services to be provided by what was left of AT &T. The original seven (7) RBOCS consisted of the following companies:

- ***Pacific Bell***
- ***Ameritech***
- ***Southwestern Bell***
- ***US West***
- ***Bell Atlantic***
- ***BellSouth***
- ***Nynex***

These companies were originally established to disperse management of responsibility across regions in the telecommunications industry. Many things have changed since the initial breakup of AT & T; though the initial purpose of breaking up AT&T was to dismantle this monopoly in the industry.

Soon changes became very prevalent in the industry. Southwestern Bell changed its name to SBC and in 1998 purchased Pacific Bell and Nevada Bell. In 1999 the newly formed SBC organization purchased Ameritech as well in an effort to rise above the competition and establish themselves as a leader in telecommunication services. In 1997 Bell Atlantic purchased Nynex and then merged with GTE to become A provider; the current largest provider of telecommunication services; as of the date this book was published. Then in the year 2000, US West merged with Qwest Communications, which is a long distance and fiber optic company. BellSouth still remains a standalone entity and encompasses local provider organizations Southern Bell and South Central Bell.

Throughout these changes, one thing has remained a constant burden for these large organizations. That being, the regulations set before them.

Federal Guidelines

An organization which must meet regulatory statutes must be constantly working to overcome the competition while following the established laws and guidelines set forth by the federal and state legislators; this was established in order to assure fair competition with smaller phone service providers and the industry.

Once AT & T was initially broken up into these different operating entities, these RBOCs were dedicated to providing the local telecommunications needs of the consumer. When establishing the RBOCs, the FCC established a checklist to provide some guidelines for these new telecommunication entities. The FCC established some minimal guidelines an RBOC needed to adhere to as a local exchange carrier. They included the following basic guidelines:

Interconnection	*Requires the RBOC to allow requesting carriers to physically link their communications networks to its network for the mutual exchange of traffic.*
Access to Unbundled Network Elements	*The RBOC must provide a connection to network elements at any technically feasible point under rates, Terms, and conditions that are just, reasonable and nondiscriminatory.*
Access to Poles, Ducts, Conduits, and Rights of Way	*The RBOC must show that competitors can obtain access to poles, ducts, conduits, and rights-of-way within reasonable time frames.*
Unbundled Local Loops	*RBOC must demonstrate that it has a concrete and specific legal obligation to furnish loops on an unbundled basis.*
Unbundled Local Transport	*Requires the RBOC to provide competitors with the transmission links on an unbundled basis that are dedicated to the use of that competitor.*
Unbundled Local Switching	*Equal access to call waiting, call forwarding and caller ID.*
911 and E911, Directory Assistance, and Operator Services.	*The RBOC must provide competing carriers with accurate and nondiscriminatory access to these services so that these*

	competitors' customers are able to reach emergency assistance.
White Pages Directory Listings	*White pages are the directory listings of telephone numbers of residences and businesses in a particular area.*
Numbering Administration	*The RBOC must provide other carriers with the same access to new NXX codes within an area code.*
Databases and Associated Signaling	*Supply Call-related databases and signaling systems that are used for billing and collection, or the transmission, routing or other provision of a telecommunications service.*
Number Portability	*Number portability enables consumers to take their phone number with them when they change local telephone companies.*
Local Dialing Parity	*The RBOC must establish that customers of another carrier are able to dial the same number of digits to make a local telephone call.*
Reciprocal Compensation	*The RBOC must compensate other local carriers for the cost of transporting and terminating a local call from the RBOC.*
Resale	*Requires the RBOC to offer other carriers all of its retail services at wholesale rates*

without unreasonable or discriminatory condition or limitations such that other carriers may resell those services to an end user.

These guidelines were established to assure adequate telecommunications interoperations within the industry. They were minimalist steps these telecom entities needed to adhere to in order to effectively operate without affecting the quality of services for the consumer. As the telecommunications infrastructure grew, it became a monopolistic entity which was in complete control of basic telecommunication services. The federal government thought that by increasing competition they would increase the availability of high-bandwidth services to meet the needs of residential and business customers. It felt that competition in the area of telecom would lead to great innovation and lead to the creation of additional jobs within the industry.

Progress in technology enabled competition to flourish due to the low cost of high speed computers and fiber optic technologies. By 1996 the FCC further addressed the monopolistic nature of telecommunications industry. It created the Telecommunications Act of 1996. Within this act the FCC removed some of the advantages that the phone providers had for almost a hundred years. It outlined many provisions which stated that competitors would be allowed to lease and resell portions of the incumbent telephone company network so they could compete without creating a brand new infrastructure. This spawned off many new smaller service providers to compete with the larger phone utilities. These new companies were referred to as competitive local exchange carriers (CLECs).

The FCC also addressed the concept of "universal service." According to the FCC, the goals of "universal service" were to directly promote the availability of quality services at reasonable and affordable rates; increase access to advanced telecommunications services throughout the Nation; advance the availability of such services to all consumers, including those in low income, rural, insular, and high cost areas at rates that were reasonably comparable to those charged in urban areas.

In addition, the 1996 Act states that all providers of telecommunications services should contribute to Federal universal service in some equitable and nondiscriminatory manner; there should be specific, predictable, and sufficient Federal and State mechanisms to preserve and advance universal service; all schools, classrooms, health care providers, and libraries should, generally, have access to advanced telecommunications services; and finally, that the Federal-State Joint Board and the Commission should determine those other principles that, consistent with the 1996 Act, are necessary to protect the public interest.

The FCC at the time viewed these provisions as a

necessary act to protect the consumer. Retention and switching of customers from the local telephone organizations to the new smaller carrier's phone service was important to the future of these new smaller phone companies. After all, most people and business already had phone service through Incumbent Local Exchange Carriers (ILECs). It was the job of the new telephone carrier to convince the consumer, that they could offer them more or better services at a lower fixed cost than the larger providers. For the first time, consumers had a choice regarding who provided their telecommunication services.

The advantage the ILECs had over the CLECs is that the older mature carriers had accumulated the embodiment of knowledge necessary to run a telephone company. They had decades of experience and a great deal of stability from an organizational perspective. The new carriers had more intimate contact with customers, they typically received more personalized attention, and, in part due to regulation, they in many situations, were able to offer products and services to the consumer at a much lower rate than their larger counterparts.

Regardless of who the consumer goes through for services, the infrastructure used is usually that of the ILEC. In fact all communications are routed from your home through the central office (CO) and is actually utilizing the ILEC's hardware when making calls or providing telecom services. Another provision stated in the act was that the Bells where not allowed to sell long distance within their regions until the FCC, on a state by state basis, are open to competition. The Justice Department grants final permission for the Bell companies to sell in-region long distance services.

UNE-P

As defined in the Telecommunications Act of 1996, telecommunications service providers must allow rivals to lease what is referred to as the *Unbundled Network Elements Platform* (UNE-P). The FCC rules like the UNE-P and other telecommunication regulations have been codified in Title 47 of the Code of Federal Regulations. They are initially published in the Federal Register. The FCC does not maintain a database of its rules nor does it print or stock copies of the rules and regulations. That task is performed by the *Government Printing Office* (GPO). After October 1st of each year, the GPO compiles all the changes, additions, and deletions to the FCC rules and publishes an updated *Code of Federal Regulations* (CFR).

The larger telecommunications service providers have packages that simplify the bundling of local, long distance and wireless services for consumers. They have streamlined operations to meet the needs of customers in various areas. However, even with these advantages, they still continue to loose customers to smaller providers. Why? Well, it is primarily because of the Telecommunications act of 1996. Larger providers are forced by Federal law to allow telecommunications rivals to utilize their networks at below cost rates which greatly affect their profit margin.

According to the FCC, and expressed in the Telecommunications Act of 1996, the larger providers, in return for being allowed into the long distance marketplace, have the duty to provide to any requesting telecommunications carrier the provision of a telecommunication service, nondiscriminatory access to network elements on an unbundled basis at any technical feasible point on rates, terms and conditions that are considered reasonable. So according to the FCC, the Telecommunications Act of 1996 provides a benefit to the larger carries as well, by allowing them to have the ability to offer long distance services.

It is a tradeoff in the realm of regulatory control and

in the view of the FCC; it was a necessary one in order to reduce the monopolistic nature of the telecommunications industry. However, it has been difficult for the larger providers to get approval from the state and federal legislation to introduce long distance in various states.

When customers change local providers from a incumbent local telephone company to a competitive local exchange carrier, the CLEC must submit change orders to the local telephone company. The CLEC has the responsibility to place the order in the proper format for effectively processing the change correctly. The incumbent telephone company in turn must in a timely and accurate manner process requests it receives from CLECs to meet these guidelines established by the FCC. If carriers do not comply with these regulations they will be fined accordingly.

It was also mandated by the FCC that all carriers will compensate each other for carrying each other's local calls. Many local service providers will sell telephone service to "Internet Service Providers" that connect users to the internet. Usually calls that the CLECs carry to ISPs originate on the incumbent local telephone company's networks. By 1999 the FCC ruled that calls to the Internet are not usually local and that reciprocal payments between carriers should be honored in this situation. By 2001, new rules regarding this issue were established. The FCC concluded that telecommunications traffic delivered to an ISP is interstate access traffic, specifically "information access," thus not subject to reciprocal compensation. However, by this time Incumbent local carriers paid billions of dollars in reciprocal fees for internet bound traffic.

Interchange Carriers

In recent years, there has been a shift in balance between local and long distance costs. Interchange carriers have to pay access fees to incumbent and competitive local telephone companies for transportation of long distance traffic between local customers. Access fees were established in an attempt to subsidize local service and assure so that rates stayed affordable for residential customers. Residential basic telephone service was also subsidized by rates businesses paid for local service and by long distance fees. However, this shifted costs for users in the form of higher monthly subscriber line charges.

As far as inter-exchange carrier compensation, the FCC has been considering eliminating all fees that carriers pay each other for interconnecting their networks. This is referred to as "bill-and-keep." Under this proposal carriers recoup the costs of originating and terminating traffic from their own customers vs. other carriers. This would eliminate reciprocal payments altogether if enacted.

Local Number Portability

Regulations specified by the FCC assure local number portability. Local number portability assures subscribers keep their telephone numbers at the same physical location when they change providers. This functionality was implemented using *Local Routing Numbers* (LRNs). Basically a 10-digit number is stored in a network database. Then when a telephone call comes in through the providers switching network, the database is accessed to retrieve this number to determine through which central office the call should be routed.

Recently wireless number portability was achieved as well. Users can now change their cellular providers without changing their cellular numbers. However, this feature is expensive to implement for service providers. It is very costly to upgrade the switches to support this new functionality. In 1999, the FCC allowed carriers to charge customers a fee to account for these incurred costs over a five year period.

In addition to the 14 RBOC requirements listed in the last chapter, the Bells must supply the following to their competition as defined in this act:

- *Reciprocal Compensation*
- *Number Portability*
- *Access to Rights-Of Way*
- *Unbundled Access*
- *Collocation of Competitive Equipment*
- *Dialing Parity(No Carrier Access Code*
- *Number Prefixes during dialing)*

In addition the act forbids carriers to automatically switch customers to another provider without their explicit authorization; commonly known as "SLAMMING." As of 1996, a third party verification scheme is used to determine if the customer actually wanted to change providers in order to meet the regulatory requirements.

Billing Strategies

Competition

In today's highly competitive telecommunications environment, it is becoming increasingly more difficult to differentiate between one provider and another. In the past only a handful of telecommunications companies existed. Now there are hundreds all fighting for a dominant presence in the industry. There are a few important strategies. One is to compete for market share by spending money in advertising costs and lowering costs to excessively low amounts to overrun the competition.

Packaging Strategies

This later approach has been successful, but people tend to switch from provider to provider and this increases support and customer maintenance costs. Another popular approach is to focus on the customer; tailor products to his/her needs, package products, use direct advertising towards a segmentation of people. For example, "the college plan," the "the family plan," etc. This requires understanding the customer using "*Customer Relationship Management*" (CRM) approaches.

This strategy many times causes a customer to become "sticky" or less likely to leave one provider for another. If a customer has multiple products tailored to how they live with one provider and they receive a high level of service, they are almost sure to remain with that provider.

Customer Focus

As telecommunication providers evolve their business, more of the focus is on a customer strategy vs. a product strategy. In other words, the industry is driving towards mass customization of product configurations for a consumer vs. mass product marketing. In order to effectively retain customers and assure customers remain "sticky," providers need to develop strategies to keep existing customers. Providers need to cater towards the needs of the consumer by offering them incentives to stay. They must evolve their IT Systems to support customer operations with respect to billing and customer care between themselves and their business alliances. They must address complex integration scenarios across affiliate business entities interfacing with ordering, billing, rating, and account management IT systems.

Generally, a telecom provider has multiple retail and wholesale sales channels. A sales channel is a point of entry into a telecom organization used to order products. A provider will, at a minimum, have the following retail channels:

- Call Center
- Self Service Web Ordering
- Integrated Voice Response

A provider's call center has a *Customer Service Representative* (CSR) on staff to answer the phone when a customer calls to order products or services. This channel provides full service. It is the doorway into the provider's enterprise. Once speaking with a CSR, the customer may be routed to any variety of departments to assist them depending on their needs.

Service Negotiation Session

The CSR will use software to order, rate, provision, and start the billing process with respect to telecom services. The actual time when the customer is on the phone ordering services with the CSR is referred to as a *Service Negotiation Session* (SNS). During a SNS the CSR may make special offers to the customer; they may "up-sell" the customer which means they may offer them a more comprehensive package. They may also transfer a customer to other support personnel if they are having problems with a service or need assistance in getting a service setup correctly. A provider's call center provides these services as well as a comprehensive list of services including the following:

- Payment Adjustments
- New Service/New Connects
- Disconnects
- Service Changes
- Sales
- Customer Support
- F & T (From and To Moves)
- CDUE (Change Due Date for Service Enablement)

Self Service Ordering

Today most providers have a self service web based ordering channel which provides the customer the ability to order telecom products and services over the internet. This sales channel allows any consumer to do anything from order products and services to billing. Even support questions are generally accepted and responded to over the internet.

These self service sales channels play an important role in the ordering process. Today, with newer wireless and handheld technologies telecom organizations are providing web based services which can be used by PDA's, wireless phones, etc.

Many times these technologies aren't reflected necessarily as a new telecom sales channel pursue. Instead they may integrate with one of the existing web channels to provide the neccessary services. For example, maybe the customer uses his or her wireless provider to connect with existing telecom services associated with the self service channel. Then using one's PDA or wireless phone the customer can order products or services.

Industry Standards

As more and more telecommunication providers begin to converge, they must continually review their current ordering and billing IT systems. They need to eliminate the use of proprietary technologies and instead pursue architectures which utilize open industry standards. From a billing IT systems perspective, the focus in today's telecom world is on integration. This includes integration between internal billing telecom systems and more importantly, integration of billing and ordering information across business partners.

In order to determine a feasible strategy, the business must perform a high level analysis of billing systems and determine feasible future end-state architectures. Then they must develop a planned phased transition strategy to meet today's business needs while addressing tomorrows anticipated needs. This requires continually performing iterative reviews of the latest ordering, customer care, and billing business processes and seeking ways to improve upon them to meet the upcoming needs of the business. The goal is always to align telecom billing and ordering strategies and establish priorities across business realms to meet every providers objective; easy integration.

Data Integration

The data integration goal of every provider is to allow affiliate product information to be integrated with the provider's billing and ordering information systems in order to offer packages which span across business partners. Also, the provider needs the ability to provide one phone bill which offers centralized billing of all affiliate packages or products.

Approaches to eliminate redundancies in business processes and logic across a provider's sales channels should be reviewed. Establishment of common functionality which can be utilized by each channel is always the less costly approach vs. repeating efforts. Centralized access and maintenance of affiliate product information also provides significant value to assure all sales channels are using up to date product information.

The core business areas to review when reviewing a billing strategy are as follows:

1. Product Management
2. Account Management
3. Rating
4. Tax and Fee Management
5. Consumer and Complex Ordering

These items in addition to centralized customer information are what providers use to reconcile ordering and billing operations between themselves and business partners. It is important to determine approaches to accelerate initiatives, reduce redundancies and provide some level of focus for the provider in order to remain competitive. If a provider does not continually evolve their business support systems, the Quality of Service (QOS) will eventually to go down and costs will be driven up in the support infrastructure.

Sales Channel Software Applications

Not only do providers need to support the ability to group products across affiliate organizations to obtain "stickiness" from their customers, they also need to reduce the amount of redundancies existing between consumer ordering applications and processes across sales channels. Software applications in different sales channels should be presenting the same business rules, offering the same promotional offers, managing business logic the same way. However, many times this is not the case due to lack of sustained integration initiatives within the business.

The importance is of centralization of processes and logic to eliminate unnecessary redundancies is crucial to minimizing a provider's expenses. Consistency and efficiency are important concepts providers need to embrace to remain competitive. There are some circumstances where a provider wants to make offers only available on a particular sales channel. For example, to reduce calls in a Call Center a provider may only make certain offers over the web. This reduces costs for the provider by making the ordering process a self service approach to provisioning and billing of telecom services.

Technology Barriers

Providers have difficult technological barriers which need to desperately be addressed. Providers need to continually review technological migration strategies to update their existing call center ordering software where future integration may be needed. The business climate has changed drastically due to technology and intense competition. Providers need to begin streamlining operations, they need to drive business rules out of their individual sales channel software and integrate these business rules instead into a common set of business services which can be common across all channels.

Software development of any new functionality related to ordering and billing for telecom services needs to be carefully managed. Anytime in a provider's environment, where software development or business processes are updated, which may affect more than one Sales

Channel, universal functionality needs to be reviewed and broken out to assure it is only developed once vs. separately for each channel. This eliminates redundancies by not reinventing the wheel and driving a provider's cost structure upward.

Providers also need to reduce complexities associated with offered products, and update the consumer ordering software systems to represent a flexible architecture which will allow functionality to be plugged-in into. Think about the "Big Mac." McDonalds did a great thing when they began offering packages for meals. It was not long before all of their competitors followed suit. Simple is always better. In the focus of customer satisfaction somewhere and at some point in the recent past companies lost site of costs vs. flexibility.

To be completely flexible, a company increases business system complexity and drives costs upward. The IT Systems and business processes are harder to use, the business rules are harder to understand, and more mistakes are common. Human error is frequent and the learning curve for new employees many times is a significant unaccounted for expense.

How can providers still be flexible yet define things simply? This is accomplished through the concept of "Bundle Management." This refers to packaging products similar to the extra value meal. A provider may offer basic phone service with a few simple features as a basic package. However, if a customer super-sizes their order the provider may throw in high speed internet as well.

This keeps things simple. Only one package needs to be defined which includes the super-sized package. This means only one pricing or rating plan, only "one" product in a sense from a billing perspective. This keeps billing and ordering business processes simple.

Bought a car lately? If you have, you probably have run across the situation where everything is considered a package. Many times taking this approach will get customer's to buy things beyond what they normally would consider. For example, think about air conditioning. Almost everyone living in a warm climate thinks it is a necessity. Therefore, they are very likely to want this feature on a new car. However, when they go to order it, it comes with the ABC package. The ABC package is $3000.00 and includes pin stripes, chrome wheels, and air conditioning. However, you only wanted air conditioning. This is a situation where the car dealer has up-sold you to a fuller set of products by forcing you to buy a package based on a common need. This is a good strategy in the telecom world as well. Packaging offers the opportunity to up-sell to a fuller set of products bringing in higher revenue for the provider.

Centralizing Common Business Logic

IT systems today have the concept of being distributed. This means an IT System may physically reside across multiple sets of hardware all across the country. Different pieces of software exist on each of these machines to process a portion of what is referred to as a "transaction." Think of a bank transaction. It includes multiple processing steps prior to being complete. This is exactly what a transaction does in an IT System. It waits until all processing steps are complete and determines whether the entire process was successful. If it wasn't, the software managing the transaction reverses the operation taking IT Systems and business processes back to their original state.

Providers today have multiple sales channels which all have software running automated operations. The software is many times distributed. For example, where common functionality is needed across sales channels such as with product information, business logic it is placed on a machine which is accessible by all channels. This software accessed in this way is many times are referred to as "middleware." Middleware is where common software business logic resides.

Providers generally place significant resources in middleware to create common services for sales channel teams. Sales Channel teams are relatively small to assure adequate synergies. These application teams need to primarily define the interface and workflow logic for individual software applications and business processes the provider needs. The majority of the difficult IT System interfacing work and telecom business logic should reside in common middleware services. Sales Channel software application teams in areas like ordering and billing would utilize these centralized services for creation of ordering and billing software to be used by different Customer Care business processes.

This eliminates redundancies between applications and assures a good telecom enterprise architecture which is robust and less costly to maintain. Today provider's many times have multiple Call Center applications with "too much" functionality embedded in them encouraging a lower quality of service than is adequate within the Call Centers.

Through the use of centralization of ordering services, product rules and continually refining account management, providers can drive the cost with supporting these multiple ordering environments out of their business. As they move towards consolidations and support of cross-affiliate product packaging strategies, providers need to focus on use of technological design patterns which assure the B2B connectivity they will need in the future to support affiliate interactions between these IT business systems.

Chapter

9

Customer Relationship Management

Intro to CRM

Customer Relationship Management (CRM) is an important business strategy which organizations today are using to drive revenue and contribute to growth. CRM does this through improved product and service targeting based on a number of environmental and customer specific attributes. This includes analysis of research based on market segments and past customer history to effectively determine a customer's needs. The goal is to assure retention of existing customers while determining future business opportunities to expand a provider's customer base at the same time.

Successful execution of this business strategy provides customers with a higher satisfaction level and assures a continued revenue stream while enforcing a business opportunity for additional sales. CRM spans across organizations, customers, processes and technology. It is not localized to any one particular category. However, automation of customer service negotiation and service ordering processes is important for successful implementation of a streamlined CRM system which fits tightly with existing telecommunications business processes.

CRM Strategies

In order to fully implement a CRM strategy, a well-designed customer knowledge repository is important. It must be tied into a software application which can be used by a *Customer Service Representative* (CSR). This is a prudent first step towards achieving a workable CRM solution. This repository could be in the form of a data warehouse or operational data store. The software application offers an opportunity ask questions to customers which are extremely valuable to the Telco organization.

Based on this information collected from consumers, providers need to implement actions in the form of product offers. This offers need to positively affect the organizations bottom line. Most telecommunication companies have captured a large amount of customer data from the multiple sales/support channels within the regions they operate. This is a powerful decision making tool when it's combined with other methods of segmentation and analysis. Companies today, in order to be effective with *Customer Relationship Management* (CRM), must focus on better serving their "best" customers. Generally, there are a few simple things companies can do to provide better customer service and improve overall satisfaction.

- *Answer the phone. People are tired of sitting in endless voice message queuing systems.*
- *Call the customer back when needed.*
- *Look for ways to create more interaction with customers.*
- *Never take customers for granted, it is too easy for them to switch service providers.*
- *Re-acquire customers. View each customer as one who is on the verge of leaving, and approach him as if he were new and had to be re-sold on your value.*
- *When contacting customers who complain, make sure it is someone knowledgeable who calls them back.*
- *Ask your customers what you could do better; they'll tell you if you listen.*

Before we go into the details of CRM, let's review the basic concepts relating to the customer provisioning services from their provider. This is a typical scenario.

When a customer orders services through their service provider, the customer service representative will generate a service order to provision the services the customer requires.

Service Order Processing

Service order processing is itself typically considered a three (3) tier processing concept. The first two tiers are referred to as the originating network, this is where the order is critiqued and modified if required, while the third tier is referred to as the completion network. Some orders will use all tiers, some will use the first and third tiers, and others will only use the third tier. When a service order is first created, it is distributed over the first tier network to work groups and mechanized business support systems required to expedite the processing of the service request before passing the final order on to the operational support systems for processing; which then provisions the network to enable the Telco switch or other service feature for the customer.

When an order is submitted to the first tier network, after reviewing the information contained in the request it sometimes requires modification due to an error in the information before the order can be processed. Once this data has been modified and verified for accuracy the order will be distributed again, this time over the second tier network.

Sometimes after the order has been submitted across both tiers, a situation will arise where the order needs to be modified yet again. For example, in response to a customer requested change. Let's say the due date needs to be changed when a customer is supposed to have a new phone line installed. This requires a change to the pending service order. When this happens, the order is changed accordingly and distributed again over the first and second tiers. This is referred to as "a-correcting" an order when this situation occurs.

When the order is ready for the actual request to be performed, it is completed over the third tier. After orders go through the third tier, they are typically distributed to the billing customer records operational support system to update the customer's information. Another part of this process which is tied to CRM, is the analytical information which is stored related to customer segmentation, personal attributes, historical ordering, etc and is part of the overall ordering and support process.

For example, if a customer places a call to their provider they are likely to get a *Customer Service Representative* (CSR) who will lookup their information. Using this information, the customer service representative can determine what products and what offers to present to the customer. How this is done is sometimes a regulatory issue. LECs must provide the same benefit to themselves as they do CLECs as indicated in the Telecommunications Act of 1996. This means they cannot use proprietary customer information to offer services unless you as a customer allow it.

Offer Management

Customers in general are rated from some type of scaling system, whether it is 1 to 4 or 1 to 10, etc… The number or level of customer determines their ability to receive offers from their provider. If you are a good customer who always pays his bills on time, then you will get preferential treatment and benefits over an individual that doesn't pay his bill on time. This ties in very much into efforts providers refer to as *offer management* (OM)

OM addresses issues relating to offers to the customer based on the type of customer they are. It will determine what bundles/products a particular customer qualifies for and at what price. OM involves many IT systems which integrate with rating systems for rating information. If offers are made across sales channels, it is important to be consistent in price.

It is important to centralize rating information in an organization. Offers made in any sales channel should be made by retrieving all rating information through one centralized IT system. Sometimes this involves a centralized IT system retrieving rates from other rating systems owned by an affiliate or business partner.

Segmentation

In order for CRM to be effective, providers must identify market segments that correspond to the circumstances in which customers find themselves when making purchasing decisions. Can they accurately theorize which products will connect with their customers? Customer segmentation (or categorization) should be based on the notion that customers use certain products to streamline particular tasks. This assists organizations in segmenting their markets to accurately mirror the way their customers use them in their every day lives.

Predictable Marketing

One of the key concepts when considering CRM implementation strategies is that you must acquire the ability to determine what customers want in a "reliable" way. This brings us to the concept of "predictable marketing" which requires a core understanding of the circumstances which lead up to why customers purchase or utilize telecommunication devices or services. When customers become aware of something that is taking up time in their daily lives, they look around for a product or service that they can use to make their life easier. This is very much the reason for integration between IT systems and telecommunications technologies.

People want automated communications in addition to having the ability to execute remote business service transactions. Essentially, we need to satisfy a requirement in our daily lives. This is how customers now experience life as we know it today through automation technology. Telecommunication organizations which target their products and business services at the circumstances in which customers find themselves, rather than at the customers themselves, are those that can launch predictably successful products.

Telecom Terms, Acronyms, and Facts

2B1Q	Two Binary One Quaternary
3-G	Third-generation mobile standard
3G	Third-Generation licenses
3G3P	3G Patent Platform Partnership
3GPP	Third-Generation Partnership Project
3D	Third generation mobile systems
4GL	Fourth Generation Language

A

A2Go	Applications2Go
AAATE	Association for the Advancement of Assistive Technology in Europe

AAC	Augmentative and Alternative Communication
ABM	Asynchronous Balances Mode (HDLC)
ACCC	Australian Competition and Consumer Commission
ACD	Automatic Call Distribution
ACEC	Advisory Commission on Electronic Commerce
ACS	Asia Cellular Satellite
ACIF	Australian Communications Industry Forum
ACITS	Advisory Committee on Information Technology Standardization
ACM	Association for Computing Machinery
ACTE	Approvals Committee for Terminal Equipr

ACTS	Advanced Communications Technologies And Services Satellite
ADA	Americans with Disabilities Act
ADL	Activities of Daily Living
ADPCM	Adaptive Differential Pulse Code Modulation
ADSL	Asymmetric Digital Subscriber Line
AER	Association for Education and Rehabilitation of the Blind and Visually Impaired
AES	Advanced Encryption Standard
AFB	American Foundation for the Blind
AIM	Advanced Informatics for Medicine
ALDICT	Access of Persons with Learning Disabilities to Information and

	Communication Technologies
ALEC	Alternate Local Exchange Carrier
AMPS	Advanced Mobile Phone System
ANI	Automatic Number Identification (ISDN)
ANSI	American National Standards Institute
AP	Application Program
APCN	Asia Pacific Cable Network
APDU	Application Protocol Data Unit
APN	Access Point Name
APNG	Asia-Pacific Networking Group
APPSN	Application Pilot for People with Special Needs
APRS	Automatic Packet Reporting System
ARCS	Astra Return Channel System
ARIN	American Registry for Internet Numbers

ARL	Akamai Resource Locator
ARM	Asynchronous Response Mode (HDLC)
ARPANET	Advanced Research Projects Agency NET
ASCII	American Standard Code for Information Interchange
ASDL	Asymmetric Digital Subscriber Line
ASI	Alternate Space Inversion
ASIC	Application Specific Integrated Circuit
ASL	American Sign Language
ASNA	Internet Assigned Numbers Authority
ASO	Address Supporting Organization
ASP	Advanced Speech Processor; Applications Service Provider
ASPIC	Application Service Provider Industry Consortium

ASR	Automatic Speech Recognition
ASP	Application Service Provider
AT	Assistive Technology
ATDM	Asynchronous Time Division Multiplexing
ATM	Asynchronous Transfer Mode or Automatic Teller Machine
ATR	Advanced Telecommunications Research (Institute)
ATRC	Adaptive Technology Research Center
ATTT	African Telecommunications Think Tank
ATUG	Australian Telecommunications Users Group
AUDETEL	Audio Description of Television for the Visually Disabled and Elderly

AVL Available

AVM Audio Visual Management

B

B2B Business-to-Business

B2C Business to Consumer

BABT British Approvals Board for Telecommunications

BAIC Barring of All Incoming Calls

BAOC Barring of All Outgoing Cells

BAP Bandwidth Allocation Protocol

BBS Bulletin Board System

BCAP Bandwidth Allocation Control Protocol

BCD Binary Coded Decimal

BDT Telecommunication Development Bureau

	of the ITU
B-ICI	Broadband Inter-carrier Interface (ATM)
BIOS	Basic Input/Output System
BIPT	Belgian Institute for Posts and Telecommunications
BIS	Business Intelligence Software
B-ISDN	Broadband Integrated Services Digital Network
BLEC	Building (or Basement) Local Exchange Carrier
BLSR	Bidirectional Line Switched Ring
BMS	Broadband Management System
BOTZ	Build Operate Transfer
BRI	Basic Rate Interface (ISDN)
BRITE	Basic Rate Interface Transmission

	Equipment
BRITE/EURAM	Basic Research in Industrial Technologies For Europe/European Research in Advanced Materials
BSC	Base Station Controller (Wireless/GSM)
BSC	Binary Synchronous Communications (IBM)
BSI	British Standards Institution
BSIC	Base Transceiver Station Identity Code
BSS	Base Switching System
BT	Business Telecommunications
BTAM	Basic Telecommunications Access Method (SNA)
BTR	British Telecom Requirement
BTS	Base Transceiver Station
BWA	Broadband Wireless Association

BWLL — Broadband Wireless Local Loop

C

CA — Certificate Authority

CALEA — Communications Assistance for Law Enforcing Agencies

CAMEL — Customized Application for Mobile Networks Enhanced Logic

CAPS — Communication and Access to Information for Persons with Special Needs

CAST — Center for Applied Special Technology

CATV — Cable Television also Community Antennae Television

CB — Citizens Band

CcTLD	Country-code Top Level Domain
CD	Call Deflection; Compact Disc
CDG	CDMA Development Organization
CDMA	Code Division Multiple Access
CDN	Content Delivery Network
CDPD	Cellular Digital Packet Data
CEC	Commission of the European Communities
CEE	Central and Eastern Europe
CEPT	Conference of European Postal and Telecommunications Administrations
CFB	Call Forward Busy
CFNR	Call Forward No Reply
CHTML	compact HTML
CIC	Carrier Identification Code

CISC	Complex Instruction Set Computer
CIX	Commercial Internet eXchange
CLEC	Competitive Local Exchange Carrier
CLI	Calling Line Identity/Identification
CLIP	Calling Line Identity Presentation
CLIR	Calling Line Identification Restriction
CLP	Cell Loss Priority (ATM)
CMCC	China Mobile Communications Corporation
CMR	Cellular Mobile Radio
CNAE	Customer Network Access Equipment
CNC	China Netcom Corporation
CNM	Customer Network Management
COLP	Connected Line Identification Presentation
COLR	Connected Line Identification Restriction

COM	Common Object Model
COMAC	Committee On Medicine And Computers
COMESA	Common Market for Eastern and Southern African States
COMETT	Community Programmer in Education and Training for Technology
	Copernicus Co-operation in Science and Technology with Central and Eastern European Countries
CORBA	Common Object Request Broker Architecture
CP	Communications Protocol
CPE	Customer Premises Equipment
CPIC	Common Programming Interface Communications
CPP	Calling Party Pays

CPS	Carrier Pre-Selection; Communications Service Providers
CREST	Scientific and Technical Research Committee
CRM	Customer Relationship Management
CRTC	Canadian Radio television and Telecommunications Commission
CSO	Committee of Senior Officers
CSDN	Circuit Switched Data Network
CSP	Converged Service Provider
CSPDN	Circuit Switched Public Data Network
CSS	Cascading Style Sheet
CT-2	Second generation Cordless Telephone, UK
CT1	European analogue cordless telephone

	system
CTI	Computer Telephony Integration
CTIA	Cellular Telecommunications Industry Association
CTR	Common Technical Regulation (EU)
CTU	Czech Telecommunications Office
CW	Call Waiting
CYTA	Cyprus Telecommunications Authority

D

D-tag	Descriptive text link
DAB	Digital Audio Broadcasting
DACS	Digital Access and Cross-connect System
DAI	Digital Audio Interface (104 kbit/s)
DAMPS	Digital Advanced Mobile Phone System

DANN	Data Networking Association
DARS	Digital Audio Radio Satellite Service
DASS	Digital Access Signaling System
DBS	Direct Broadcast Service
DCC	Data Communication Channel
DCN	Data Communications Network
DCS	Digital Cellular System
DECT	Digital European Cordless Telecommunications
DES	Data Encryption Standard
DFID	Department For International Development
DGPS	Differential Global Positioning System
DHTML	Dynamic HTML
DIS	Draft International Standards

DNS	Domain Name System
DNSO	Domain Name Supporting Organization
DOCSIS	Data Over Cable Service Interface Specification
DOM	Document Object Specification
DPL	Digital Power Line
DRB	Digital Radio Broadcasters
DS	Direct Spread
DSL	Digital Subscriber Line
DSLAM	DSL Access Multiplexer
DSRR	Digital Short Range Radio
DTD	Document Type Definition file
DTH	Direct-To-Home
DVB	Direct digital Video Broadcasting
DWDM	Dense Wave (or Wavelength) Division

Multiplexing

E

E&D Elderly and Disabled

EARN European Academic Research Network

e-B2B electronic business-to-business

EBU European Broadcasting Union

EC European Community, European Commission

ECART European Conference on Advances in Rehabilitation Technology

ECDC Electronic Commerce in Developing Countries

ECMA European Computer Manufacturers Association

ECOWAS Economic Community of West African

	States
eCRM	electronic Customer Relation Management
ECTA	European Competitive Telecommunications Association; Eastern Caribbean Telecommunications Authority
ECUMN	European Conference on Universal Multi-Service Networks
EDF	European Disability Forum
EDGE	Enhanced Data for GSM Evolution; Enhanced Data rates for Global Evolution
EDI	Electronic Data Interchange
EEC	European Economic Community
EEMA	European Forum for Electronic Business
EESSI	European Electronic Signature Standardization Initiative
EFEI	European Federation of Electronic Industries

EFF	Electronic Frontier Foundation
EFTA	European Free Trade Association
EFTPOS	Electronic Funds Transfer Point of Sale
EHIMA	European Hearing Instrument Manufacturers Association
EIA	Electronic Industries Association
EIDD	European Institute for Design and Disability
EIF	European Internet Foundation
EIIA	European Information Industry Association
EITAAC	Electronic Information Technology Access Advisory Committee
EMC	Electromagnetic Compatibility
EMI	Electromagnetic Interference

EN	European Norm (Term within CEN and CENELEC)
ENSOETSI	National Standardization Organizations (ETSI)
ENUM	Electronic Numbering
EOTC	European Organization for Testing and Certification
E-OTD	Enhanced Observed Time Difference
EP	European Parliament
ERO	European Radio communications Office
ERP	Enterprise Resource Planning
ERPS	Enhanced Reference Picture Selection
ES	European Standard or End Systems
ESB	European Standardization Board
ESC	European Standardization Council

ESO	European Standardization Organization
ESPRIT	European Strategic Program for Research and Development
ETAN	European Technology Assessment Network
ETIS	European Telecommunications Informatics Services
ETNO	European public Telecommunications Network Operators association
ETO	European Telecommunications Organization
ETP	European Telecommunications Platform
ETR	ETSI Technical Report
ETT	European Telecommunications and Technology

ETS	European Telecommunications Standard
ETSI	European Telecommunications Standards Institute
EU	European Union
EUREKA	European Research Co-operation Agency
EURESCOM	European Institute for Research and Strategic Studies in Telecommunications
EuroISPA	European Internet Service Providers Association
EVA	Economic Value Added
EVCA	European Venture Capital Association
EVUA	European Virtual Private Network Users Association

F

FCC	Federal Communications Commission
FEDMA	Federation of European Direct Marketing
FDMA	Frequency Division Multiple Access
FEI	Federation of the Electronics Industry
FIPS	Federal Information Processing Standards
FLAG	Fiber optic Link Around the Globe
FMC	Fixed Mobile Convergence
FP6	the EU's 6th Framework research Program
FSS	Fixed Satellite Service
FS-VDSL	Full Service-Very High Speed Digital Subscriber Line
FTP	File Transfer Protocol (IETF)
FTC	Federal Trade Commission
FTSE	Financial Times-Stock Exchange
FWA	Fixed Wireless Access

G

GAC	(ICANN's) Government Advisory Committee
GAIT	GSM MAP Network/ANSI-41 Interoperability Team
GAM	Global Account Management
GAN	Global Area Network
GATT	General Agreement on Tariffs and Trade
GBDe	Global Business Dialog on electronic commerce
GEO	Geosynchronous Earth Orbit
GGRF	GSM Global Roaming Forum
GGSN	Gateway GPRS Support Node
GIS	Geographical Information System

GMCF	Global Mobile Commerce Forum
GMPCS	Global Mobile Personal Communication Systems (or Services)
GNS	Global Network Strategies
GNX	GlobalNet eXchange
GPRS	General Packet Radio Service
GPS	Global Positioning System
GPRS	General Packet Radio Services
GPSR	General Packet Switched Radio
GSA	GSM System Area
GRX	GPRS Roaming Exchange
GSM	Global System for Mobile Telecommunications
GSP	Global Service Provider
gTLD	generic Top-Level Domain

GTS	Global TeleSystems Group
GUIB	Graphical User Interfaces for Blind people

H

HAC	Hearing Aid Compatible
HAPS	High Altitude Platforms
HDLC	High-level Data Link Control (ISO)
HDML	Handheld Devices Markup Language
HDSL	High-rate Digital Subscriber Loop
HDTV	High Definition Television
HDVS	High Definition Video System
HKTUG	Hong Kong Telecommunications Users Group
HSCD	High Speed Circuit Switched Data
HSCSD	High Speed Circuit Switched Data
HTTP	Hypertext Transfer Protocol (IETF)

I

IAB	Internet Architecture Board
IAHC	Internet Ad Hoc Committee
IANA	Internet Assigned Numbers Authority
IASP	Internet Access Service Provider
IB	International Bureau (of the FCC)
IBC	Integrated Broadband Communication
I-CAN	Integrated Customer Access Network
ICAN	Integrated Control of ´All Needs
ICANN	Internet Corporation for Assigned Names and Numbers
ICC	International Chamber of Commerce
ICD	International Classification of Diseases
ICT	Information and Communications

	Technology
ICTA	International Commission on Technology and Accessibility
IDN	Integrated Digital Network
IDP	International Data Post
iDSN	internationalized Domain Name System
IEC	International Electro technical Commission
IEEE	Institute of Electrical and Electronics Engineers
IEGMP	Independent Expert Group on Mobile Phones
IETF	Internet Engineering Task Force
IFHOH	International Federation of Hard of Hearing People
ILEC	Incumbent Local Exchange Carrier
ILETS	International Law Enforcement Telecommunications Seminar

IM	Instant Messaging
IMAP	Internet Managed Application Provider
IMP	Interface Message Processor
IMS	Information Management System
IMSO	International Maritime Safety Organization
IMT	International Mobile Telephony
IMTC	International Multimedia Teleconferencing Consortium
IN	Intelligent Networks
INCLUDE	Inclusion of Disabled and Elderly in Telematics
INMARSAT	International Marine Satellite Organization
INMS	International Network Management System

INTELSAT	International Telecommunications Satellite Organization
INTUG	International Telecommunications Users Group
IOP	Input/Output Processor
I/P	Internet Protocol
IP	Internet telephony Providers; Internet Protocol; Intellectual Property
IPDR	Internet Protocol Detail Record
IPLC	International Private Leased Circuit
IPO	Initial Public Offering
IPR	Intellectual property rights
IPSNI	Integration of People with Special Needs by IBC
IPv4	Internet Protocol version 4
Ipv6	Internet Protocol version 6

IP VPN	Internet Protocol Virtual Private Network
IrDA	Infrared Data Association
IRG	Independent Regulators Group
IRU	Indefeasible Right of Use
ISC	International Switching Centre
ISDN	Integrated Services Digital Network
ISF	International Sensitivity Forum
ISIS	Information Society Initiative for Standardization
ISM	Industrial Scientific-Medical (frequency band)
ISO	International Organization for Standardization
ISOC	Internet Society
ISP	Internet Service Provider

ISPA	Internet Service Providers Association
IST	Information Society Technologies
ISTAG	IST Advisory Group
ISVs	Independent Software Vendors
IT	Information Technology
ITM	Information/Transaction Machine
ITMC	International Multimedia Teleconferencing Consortium
ITS	Intelligent Transport Systems
ITSP	Internet Telephony Service Provider
IT&T	Information Technology and Telecommunications
ITU	International Telecommunications Union
ITU-R	ITU Radio communication standardization sector

ITU-S	ITU Standardization bureau
ITU-T	ITU Telecommunication Standardization Sector
IUDD	Infrastructure and Urban Development Department
iVAD	integrated Voice and Data
IVR	Interactive Voice Response

J

JAIN	Java APIs for Integrated Networks
JCG	Joint Coordination Group
JPEG	Joint Photo graphics Expert Group
JTC	Jordan Telecommunications Company
JUS	Japan-U.S. Cable Network

K

KM	Knowledge Management
KPN	Telecom formerly PTT Telecom BV

L

LAN	Local Area Network
LANS	Local Area Network Services
LAP	Link Access Protocol
LAS-CDMA	Large Area Synchronized Code Division Multiple Access
LCN	Local Communication Network
LDAP	Lightweight Directory Access Protocol
LDS	Location Dependent Services
LEC	Local Exchange Carrier
LEO	Low Earth Orbit
LIMS	Laboratory Information Management System

LMDS	Local Multipoint Distribution System
LRIC	Long Run Incremental Cost

M

MAC	Media Access Control
MADM	Multiple Add/Drop Multiplexer
MAN	Metropolitan Area Network
MAP	Management Application Protocol
MC	Multi-Carrier
MCPA	Mulit Carrier Power Amplifier
MDA	Mobile Data Association
MDD	Medical Device Directive
MDF	Main Distribution Frame
MDU	Multiple Dwelling Unit

MIME	Multipurpose Internet Mail Extensions (IETF)
MLA	Mutual Legal Assistance
MMAC	Multimedia Mobile Access Communications
MNO	Mobile Network Operators
MoU	Memorandum of Understanding
MOWLAN	Mobile Operator Wireless LAN
MPEG	Motion Picture Experts Group (ISO)
MPLS	Multi-Protocol Label Switching
MSDL	MPEG-4 Syntactic Description Language
MSN	Multiple Subscriber Number; Microsoft Network
MSO	Multiple Systems Operator
MSS	Mobile Satellite Service; Mobile Satellite Systems
MTS	Mobile Telecommunications Services

MVA	Market Value Added
MVNO	Mobile Virtual Network Operator

N

NAFTA	North American Free Trade Agreement
NAF	National Arbitration Forum
NAPTR	Naming Authority Pointer
NASSCOM	National Association of Software and Service Companies
NBAR	Network Based Application Recognition
NCAM	National Center for Accessible Media
NCTA	National Cable Television Association
NNTH	Nordic Forum for Telecommunication and Disab
NGI	Next Generation Integration
NGO	Non Governmental Organizations

NLSP	NetWare Link Services Protocol
NMSI	National Mobile Station Identification number
NMT	Nordic Mobile Telephone
NRA	National Regulatory Authority
NSH	Nor Cooperation on Disability
NSI	Network Solutions Incorporated
NSO	National Standards Organizations
NSP	Network Service Protocol
NTII	National Tele-Immersion Initiative
NTIA	National Telecommunications and Information Administration (USA)
NTS	Number Translation Services
NTSC	National Television System Committee

O

OCN	Operational Carrier Number
ODSI	Optical Domain Service Interconnect
OECD	Organization for Economic Cooperation and Development
OECS	Organization of Eastern Caribbean States
OEM	Original Equipment Manufacturer
OFDM	Orthogonal Frequency Division Multiplexing
OFTEL	Office of Telecommunications (UK)
OHG	Operators Harmonization Group
OIF	Optical Internet Forum
OLED	Organic Light Emitting Diode
OMTN	Operation and Maintenance for Telecoms Networks
ONA	Open Network Architecture

ONP	Open Network Provision
OOPS	Object-Oriented Programming System
OPEN	Orientation by Personal Electronic Navigation
ORBIT	Open Market Reorganization for the Betterment of International Telecommunications
OSCE	Organization for Security and Cooperation in Europe
OSI	Open Systems Interconnect (ion)
OSRP	Optical Switching and Routing Protocol
OSS	Operational Support Systems; Operations Systems and Software

P

PABX	Private Automatic Branch Exchange
PAC	Program Advisory Committee (ETSI)
PACS	Personal Access Communications System
PAIX	Palo Alto Internet Exchange
PATU	Pan African Telecommunication Union
PBX	Private Branch Exchange
PCCW	Pacific Century Cyber Works
PCG	Pacific Century Group
PCI	Programmable Communications Interface
PCM	Pulse Code Modulation
PCN	Personal Communication Network
PCS	Personal Communications Services
PDA	Personal Digital Assistant
PDF	Portable Document Format
PGP	Pretty Good Privacy

PKI	Public Key Infrastructure
PMR	Private Mobile Radio
POP	Point of Pressure
POTS	Plain Old Telephone Service
POS	Point Of Sale
POTS	Plain Ordinary Telephone Service
PPM	Price per Minute
PSDN	Packet Switched Data Network
PS	Packet Switching
PSN	People with Special Needs; Public Switched Network
PSO	Protocol Supporting Organization
PSPDN	Packet Switched Public Data Network
PSTN	Public Switched Telephone Network
PTC	Pacific Telecommunications Conference

PTCL Pakistani Telecommunications Corporation Limited

PTO Public Telecommunication Operator

PTS Public Telecommunications System

Q

QoS Quality of Service

R

RACE Research and development in Advanced Communications technologies in Europe

RACF Radio Access Control Function

RAVE Remote Access Verification Environment

RBOC	Regional Bell Operating Company
RDS	Radio Data System
RIAA	Recording Industry Association of America
RIP	Regulatory Investigative Powers
RIR	Regional Internet Registries
RISC	Reduced Instruction Set Computer
RLSD	Receive Line Signal Detector
RSS	Residential Service Site
RT	Rehabilitation Technology
RTD	Research and Technical Development
RTF	Radio Terminal Function

S

SADC	Southern Africa Development Community

SAR	Specific Absorption Rate
SATRA	South African Telecommunications Regulatory Agency
SCSI	Small Computer Systems Interface
SCVF	Single Channel Voice Frequency
SDMI	Secure Digital Music Initiative
SDH	Synchronous Digital Hierarchy-based
SDN	Switched Digital Network
SDRU	Sensory Disabilities Research Unit
SDSL	Symmetric Digital Subscriber Line
SESAT	Siberian European Satellite
SET	Secure Electronic Transfer
SFA	Systems Financial Analysis
SGML	Standard Generalized Markup Language
SGSN	Serving GPRS Support Node

SHDSL Symmetrical High-bit-rate DSL

SID System ID number

SIM Set Initialization Mode (HDLC); Subscriber Identity Module; Subscriber Information Module

SIP Session Initiation Protocol; Smart Image Processor

SLA Service Level Agreement

SLC Subscriber Line Charge

SLM Single Longitudinal Mode

SMART Strategy for Mobile Advanced Radio Telecommunications

SMDS Switched Multi-megabit Data Service (Bellcore)

SME Small and Medium-sized Enterprises

SMIL	Synchronized Multimedia Integration Language
SMP	Significant Market Power
SMS	Short Message Service
SMTP	Simple Mail Transfer Protocol (IETF)
SNA	Synchronous Network Architecture; Systems Network Architecture
SOGIT	Senior Officials Group for Information Technology
SOHO	Small Office/Home Office
SONET	Synchronous Optical Network
SPARC	Scalable Performance Architecture; Scholarly Publishing and Academic Resources Coalition
SPP	Sequenced Packet Protocol (NetWare)

SPRINT	Strategic Program for Innovation and Technology Transfer in Europe
SSBSC	Single Sideband Suppressed Carrier
SSL	Secure Socket Level (data scrambler)
STAPL	Standard Test and Programming Language
STB	Set Top Box
STD	Subscriber Trunk Dialing
STM	Synchronous Transfer Mode
STRIDE	Science and Technology Research into Innovative Developments in Europe
SWIFT	Society for Worldwide Interbank Financial Telecommunications (consortium of banks)

T

TAC	Terminal Access Controller

TACD	Transatlantic Consumer Dialog
TAG	Technical Advisory Group; Telecommunications Action Group
TAIS	Total Aircraft Information System
TCAP	Transaction Capabilities Application Part
TCP/IP	Transmission Control Protocol/Internet Protocol
TC HF	Technical Committee Human Factor (ETSI)
TCIF	Telecommunications Industry Forum
TCP	Transmission Control Protocol
TCT	Toll Connecting Trunk
TDCC	Transportation Data Coordination Committee
TD-CDMA	Time Division-code CDMA

TDA	Time Distance of Arrival
TDD	Telecommunication Device for the Deaf
	Telecommunication Display Device
	Time Division Duplex
TDM	Time Division Multiplexing
TDMA	Time Division Multiple Access
TDS	Time Division Switching
TER	Technical basis for regulation
TERENA	Trans-European Research and Education Networking Association
TIA	Telecommunications Industry Association
TINA-C	Telecommunications Intelligent Networking Architecture Consortium
TIST	Telecommunication Information Science Technology System

TLD	Top Level Domain
TMA	Telecommunications Managers' Association
TR	Token Ring
TRAC	Technical Recommendations Application Committee
TSAG	Telecommunication Standardization Advisory Group
TSB	Telecommunications Standardization Board (ITU)
TTP	Trusted Third Party
TTY	Teletype (writer)
TUG	Telecommunication User Group

U

UAS	Universal Access System
DU	Universal Design
UDM	Unbundled Network Elements
UDRP	Uniform Domain Resolution Policy (or Panel)
UM	Unified Messaging
UMTS	Universal Mobile Telecommunications System
UN/CE	United Nations Commission for Europe
UNL	Universal Networking Language
UPU	Universal Postal Union
URETS	Uniform Rules for Electronic Trade and Settlement
URI	Uniform Resource Identifier
URL	Uniform Resource Locator (WWW)
USB	Universal Serial Bus

USI	Universal Services Issues
USISPA	US ISP Alliance
USTR	U.S. Trade Representative
UUCP	Unix-to-Unix Copy Protocol
UWCC	Universal Wireless Communications Consortium

V

VA	Validation Authority
VAN	Value Added Network
VDSL	Very High-speed Digital Subscriber Line
VDT	Visual Display Terminal
VDU	Visual Display Unit
VISP	Virtual Internet Service Provider
VLIW	Very Long Instruction Word

VMO	Virtual Mobile Operators
VMS	Voice Messaging System
VOD	Video On Demand
VoDSL	Voice-over-DSL
VoIP	Voice-over Internet Protocol
VON	Voice Over the Net
VPN	Virtual Private Network
VPOP	Virtual Point-Of-Preference
VSAT	Very Small Aperture Terminal

W

W3C	World Wide Web Consortium
WAB	Web Accessibility for the Blind
WAI	Web Accessibility Initiative
WAIS	Wide Area Information Server

WAP	Wireless Application Protocol
W-CDMA	Wide-band CDMA
WAI	Web Accessibility Initiative
WAP	Wireless Application Protocol
WARC	World Administrative Radio Conference
WASP	Wireless ASP
WATS	Wide Area Telecommunications Service
WBU	World Blind Union
W-CDMA	Wideband Code Division Multiple Access
WDM	Wave Division Multiplexing
WFD	World Federation of the Deaf
WHO	World Health Organization
WIM	Wireless Identification Module
WIMP	Windows, Icons, Mouse, and Pull-down menus

WIPO	World Intellectual Property Organization
WIVA	Wireless Internet Venture Association
WLAN	Wireless Local Area Network
WLL	Wireless Local Loop
WLS	Wireless Location Services
WP	Working Paper
WPASET	Web Page Accessibility Self-Evaluation Test
WML	Wireless Markup Language
WP4	Work Program 4
WRC	World Radio communications Conference (branch of ITU)
WTC	World Trade Center
WTCA	World Trade Center Association
WTSC	World Telecommunication Standardization Conference (ITU)

WTO World Trade Organization

X

xDSL x-Type Digital Subscriber Line

XML Extended Mark-up Language

Bibliography

Boettinger, H. M., The Telephone Book: Bell, Watson, Vail and American Life, 1876-1983, Stearn Publishers, New York, 1984.

Brooks, J., Telephone. The First Hundred Years, Harper & Row, New York, 1976.

AT&T, Events in Telecommunications History, AT&T, New York, 1983.

Bruce, R. V., Bell. Alexander Graham Bell and the Conquest of Solitude, Little, Brown, Boston, 1973.

Gamet, R. W., The Telephone Enterprise: The Evolution of the Bell System's Horizontal Structure, Johns Hopkins University Press, Baltimore, 1985.

Henck, R W., and Strassburg, B., A Slippery Slope. The Long Road to the Breakup of AT&T, Greenwood Press, Westport, CT, 1988.

Paine, A. B., Theodore N. Vail, A Biography, Harper & Brothers, New York, 1931.

Pier, A. S., Forbes- Telephone Pioneer, Dodd, Mead, New York, 1953.

Schlesinger, L. A. et al.. Chronicles of Corporate Change. Management Lessons from AT&T and Its Offspring, Lexington Books, Lexington, MA, 1987.

Smith, G. D., The Anatomy of a Business Decision: Bell, Western Electric and the Origins of Vertical Integration, Johns Hopkins University Press, Baltimore, 1985.

Stehman, J. W., The Financial History of the American Telephone & Telegraph Company, Houghton, Boston, 1925.

Sterling, C. H. et al., eds., Decision to Divest. Major Documents in US. v. AT&T, Communications Press, Washington, DC, 1986.

Stone, A., Wrong Number.- The Breakup of AT&T, Basic Books, New York, 1989.

Temin, P., with Galambos, L., The Fall of the Bell System, Cambridge University Press, New York, 1987.

Todd. K. P., Jr. (compiler), A Capsule History of the Bell

System, AT&T, New York, 1979.

Tunstall, W. B., Disconnecting Parties. Managing the Bell System - An Inside View, McGraw-Hill, New York, 1985.

Greenberg, CRM at the Speed of Light, McGraw-Hill, New York, 2002

Greenberg, CRM at the Speed of Light, McGraw-Hill, New York, 2002

M.S. Mastel Telecom Audit, McGraw-Hill, New York, 2003

Printed in the United States
36476LVS00002B/69